This report contains the collective views of an international group of experts and does not necessarily represent the decisions or the stated policy of the United Nations Environment Programme, the International Labour Organisation, or the World Health Organization.

Environmental Health Criteria 120

PARTIALLY HALOGENATED CHLOROFLUOROCARBONS (METHANE DERIVATIVES)

Published under the joint sponsorship of the United Nations Environment Programme, the International Labour Organisation, and the World Health Organization

Draft prepared by Professor D. Beritic-Stahuljak and Professor F. Valic, University of Zagreb, Yugoslavia, using texts made available by Dr D.S. Mayer, Hoechst AG, Frankfurt am Main, Germany and by Dr I.C. Peterson and Dr G.D. Wade, ICI Central Toxicological Laboratory, Macclesfield, United Kingdom

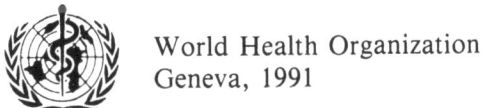

World Health Organization
Geneva, 1991

The **International Programme on Chemical Safety (IPCS)** is a joint venture of the United Nations Environment Programme, the International Labour Organisation, and the World Health Organization. The main objective of the IPCS is to carry out and disseminate evaluations of the effects of chemicals on human health and the quality of the environment. Supporting activities include the development of epidemiological, experimental laboratory, and risk-assessment methods that could produce internationally comparable results, and the development of manpower in the field of toxicology. Other activities carried out by the IPCS include the development of know-how for coping with chemical accidents, coordination of laboratory testing and epidemiological studies, and promotion of research on the mechanisms of the biological action of chemicals.

WHO Library Cataloguing in Publication Data

Partially halogenated chlorofluorocarbons (methane derivatives).

(Environmental health criteria ; 126)

1.Freons - adverse effects 2.Freons - toxicity 3.Environmental exposure I.Series

ISBN 92 4 157126 8   (NLM Classification: QV 633)
ISSN 0250-863X

©World Health Organization 1991

Publications of the World Health Organization enjoy copyright protection in accordance with the provisions of Protocol 2 of the Universal Copyright Convention. For rights of reproduction or translation of WHO publications, in part or *in toto*, application should be made to the Office of Publications, World Health Organization, Geneva, Switzerland. The World Health Organization welcomes such applications.

The designations employed and the presentation of the material in this publication do not imply the expression of any opinion whatsoever on the part of the Secretariat of the World Health Organization concerning the legal status of any country, territory, city, or area or of its authorities, or concerning the delimitation of its frontiers or boundaries.

The mention of specific companies or of certain manufacturers' products does not imply that they are endorsed or recommended by the World Health Organization in preference to others of a similar nature that are not mentioned. Errors and omissions excepted, the names of proprietary products are distinguished by initial capital letters.

PRINTED IN FINLAND
91/8922 — Vammala — 5500

CONTENTS

ENVIRONMENTAL HEALTH CRITERIA FOR PARTIALLY
HALOGENATED CHLOROFLUOROCARBONS (METHANE
DERIVATIVES)

INTRODUCTION 12

1. SUMMARY 15

    1.1 Identity, physical and chemical properties,
and analytical methods 15
    1.2 Sources of human and environmental exposure 15
    1.3 Environmental transport, distribution, and
transformation 16
    1.4 Environmental levels and human exposure 16
    1.5 Kinetics and metabolism in laboratory animals
and humans 16
    1.6 Effects on laboratory mammals and *in vitro* test
systems 17
    1.7 Effects on humans 19
    1.8 Effects on other organisms in the laboratory
and field 19
    1.9 Evaluation and conclusions 20

2. IDENTITY, PHYSICAL AND CHEMICAL PROPERTIES,
AND ANALYTICAL METHODS 22

    2.1 Identity 22
    2.2 Physical and chemical properties 22
    2.3 Conversion factors 22
    2.4 Analytical methods 22

3. SOURCES OF HUMAN AND ENVIRONMENTAL
EXPOSURE 26

    3.1 Natural occurrence 26
    3.2 Anthropogenic sources 26
        3.2.1 Production levels 26
        3.2.2 Manufacturing processes 26
        3.2.3 Loss during disposal, transport, storage,
and accidents 27
    3.3 Use patterns 27
        3.3.1 Major uses 27

	3.3.2 Releases during use: controlled or uncontrolled	28
4.	ENVIRONMENTAL TRANSPORT, DISTRIBUTION, AND TRANSFORMATION	29
4.1	Biodegradation and bioaccumulation	29
4.2	Environmental transformation and interaction with other environmental factors	29
5.	ENVIRONMENTAL LEVELS AND HUMAN EXPOSURE	32
5.1	Environmental levels	32
	5.1.1 Air	32
	5.1.2 Water	33
	5.1.3 Food and other edible products	33
5.2	General population exposure	33
5.3	Occupational exposure	34
6.	KINETICS AND METABOLISM IN LABORATORY ANIMALS AND HUMANS	35
6.1	Animal studies	35
	6.1.1 Absorption	35
	6.1.2 Distribution	36
	6.1.3 Metabolic transformation	36
	6.1.4 Elimination	38
6.2	Human studies	39
	6.2.1 Absorption and elimination	39
	6.2.2 Distribution	40
7.	EFFECTS ON LABORATORY MAMMALS AND *IN VITRO* TEST SYSTEMS	41
7.1	Single exposure	41
	7.1.1 Acute oral toxicity	41
	7.1.2 Acute inhalation toxicity	41
7.2	Short-term inhalation exposure	42
	7.2.1 HCFC 21	43
	7.2.2 HCFC 22	45
	7.2.3 Mixed exposure	46
7.3	Skin and eye irritation; sensitization	46
	7.3.1 Skin irritation	46
	7.3.2 Eye irritation	47

		7.3.3 Skin sensitization	48
	7.4	Long-term inhalation exposure	48
	7.5	Reproduction, embryotoxicity, and teratogenicity	49
		7.5.1 Reproduction	49
		7.5.2 Embryotoxicity and teratogenicity	50
		7.5.2.1 HCFC 21	50
		7.5.2.2 HCFC 22	51
	7.6	Mutagenicity	54
		7.6.1 HCFC 21	54
		7.6.2 HCFC 22	54
	7.7	Carcinogenicity	58
	7.8	Special studies - cardiovascular and respiratory effects	60
		7.8.1 HCFC 21	61
		7.8.2 HCFC 22	62

8. EFFECTS ON HUMANS 65

   8.1 General population exposure 65
      8.1.1 Accidents 65
      8.1.2 Controlled human studies 65
   8.2 Occupational exposure 65

9. EFFECTS ON OTHER ORGANISMS IN THE LABORATORY AND FIELD 68

10. EVALUATION OF HUMAN HEALTH RISKS AND EFFECTS ON THE ENVIRONMENT 69

   10.1 Evaluation of human health risks 69
      10.1.1 Direct health effects resulting from exposure to partially halogenated chlorofluorocarbons 69
      10.1.2 Health effects expected from a depletion of stratospheric ozone by partially halogenated chlorofluorocarbons 71
   10.2 Effects on the environment 71

11. CONCLUSIONS AND RECOMMENDATIONS FOR PROTECTION OF HUMAN HEALTH AND THE ENVIRONMENT 72

   11.1 Conclusions 72

11.2	Recommendations for protection of human health and the environment	72
12.	FURTHER RESEARCH	74
13.	PREVIOUS EVALUATIONS BY INTERNATIONAL BODIES	75
	REFERENCES	76
	RESUME	86
	RESUMEN	92

WHO TASK GROUP ON ENVIRONMENTAL HEALTH CRITERIA FOR PARTIALLY HALOGENATED CHLOROFLUOROCARBONS (METHANE DERIVATIVES)

Members

Professor D. Beritić-Stahuljak, Andrija Štampar School of Public Health, University of Zagreb, Zagreb, Yugoslavia

Dr B. Gilbert, Technology Development Company (CODETEC), Cidade Universitaria, Campinas, Brazil (*Joint Rapporteur*)

Professor H.A. Greim, Institute of Toxicology, Association for Radiation and Environmental Research, Neuherberg, Germany (*Chairman*)

Dr H. Illing, Health and Safety Executive, Merseyside, United Kingdom

Dr W. Jameson, Office of the Senior Scientific Advisor to the Director, National Institute of Environmental Health Sciences, Research Triangle Park, North Carolina, USA

Dr H. Kraus, Chemicals Hazardous to the Environment, Federal Ministry for the Environment, Nature Conservation and Nuclear Safety, Bonn, Germany

Dr J. Sokal, Department of Toxicity Evaluation, Institute of Occupational Medicine, Lodz, Poland

Dr V. Vu, Oncology Branch, Office of Toxic Substances, US Environmental Protection Agency, Washington, DC, USA

Observers

Dr D.S. Mayer, Department of Toxicology, Hoechst AG, Frankfurt am Main, Germany

Dr H. Trochimowicz, E.I. Du Pont de Nemours & Co., Haskell Laboratory for Toxicology and Industrial Medicine, Newark, Delaware, USA

Secretariat

Professor F. Valić, Consultant, IPCS, World Health Organization, Geneva, Switzerland, *also* Vice-Rector, University of Zagreb, Zagreb, Yugoslavia (*Responsible Officer and Secretary*)

Dr S. Swierenga, Health and Welfare Canada, Ottawa, Canada, *also* Representative of the International Agency for Research on Cancer, Lyon, France (*Joint Rapporteur*)

NOTE TO READERS OF THE CRITERIA MONOGRAPHS

Every effort has been made to present information in the criteria monographs as accurately as possible without unduly delaying their publication. In the interest of all users of the environmental health criteria monographs, readers are kindly requested to communicate any errors that may have occurred to the Manager of the International Programme on Chemical Safety, World Health Organization, Geneva, Switzerland, in order that they may be included in corrigenda, which will appear in subsequent volumes.

* * *

A detailed data profile and a legal file can be obtained from the International Register of Potentially Toxic Chemicals, Palais des Nations, 1211 Geneva 10, Switzerland (Telephone No. 7988400 or 7985850).

ENVIRONMENTAL HEALTH CRITERIA FOR PARTIALLY HALOGENATED CHLOROFLUOROCARBONS (METHANE DERIVATIVES)

A Task Group on Environmental Health Criteria for Partially Halogenated Chlorofluorocarbons (Methane Derivatives) met at the Institute of Toxicology, Neuherberg, Germany, from 17 to 21 December 1990. Professor H.A. Greim opened the meeting on behalf of the host institute. Dr H. Kraus spoke on behalf of the Federal Government, which sponsored the meeting. Professor F. Valić welcomed the members on behalf of the three cooperating organizations of the IPCS (UNEP/ILO/WHO). The Task Group reviewed and revised the draft criteria document, made an evaluation of the direct and indirect risks for human health from exposure to the partially halogenated chlorofluorocarbons reviewed, and made recommendations for health protection and further research.

The first draft on HCFC 21 was prepared by Dr D.S. Mayer (Department of Toxicology, Hoechst AG, Frankfurt am Main, Germany) and on HCFC 22 by Dr I.C. Peterson and Dr J.D. Wade (ICI Central Toxicology Laboratory, Macclesfield, United Kingdom). The second draft of the monograph was prepared by Professors D. Beritić-Stahuljak and F. Valić.

Professor F. Valić was responsible for the overall scientific content, and Dr P.G. Jenkins, IPCS, for the technical editing of the monograph.

* * *

Financial support for the Task Group meeting was provided by the Ministry for the Environment, Nature Conservation and Nuclear Safety, Germany, which also generously supported the cost of printing this monograph.

ABBREVIATIONS

ALAT	alanine aminotransferase
ASAT	aspartate aminotransferase
CNS	central nervous system
ECG	electrocardiogram
EEC	European Economic Community
FSH	follicle stimulating hormone
GWP	global-warming potential
HCFC	hydrochlorofluorocarbon
ip	intraperitoneal
LH	luteinizing hormone
ODP	ozone-depletion potential
TLV	threshold limit value
UNEP	United Nations Environment Programme

INTRODUCTION

Chlorofluorocarbons were developed as refrigerants some 60 years ago. However, their application soon significantly diversified, owing to their properties of non-flammability, chemical and thermal stability, and generally low toxicity. They are now used as blowing agents in foam insulation production, as propellants in aerosols, as cleaning agents of metals and electronic components, and to a lesser extent as chemical intermediates. Their current production is more than 1 000 000 tonnes per year with a market value estimated to be close to US$ 1.5 billion.

Chlorofluorocarbons are very stable compounds, which remain intact in the air, releasing chlorine only when they reach the stratosphere. The active chlorine destroys ozone molecules, thus depleting the ozone layer, which is a natural barrier to ultraviolet radiation potentially harmful to human health and the environment.

The growing global concern over this effect resulted in the development of the Vienna Convention for the Protection of the Ozone Layer, adopted in March 1985, and its "Montreal Protocol on Substances that Deplete the Ozone Layer", signed by 24 countries in September 1987. The agreement required a freeze in the production and use of the fully halogenated chlorofluorocarbons 11, 12, 113, 114, and 115 at 1986 levels by mid-1989, a 20% reduction in their use from 1 July 1993, and a further 30% reduction from 1 July 1998. The Protocol has been in effect since January 1989; 67 countries and the European Economic Community had signed the Protocol by July 1989. As a further development, the Helsinki Declaration, a non-binding agreement of April 1989, called for a total phase-out of the fully halogenated chlorofluorocarbons. The European Economic Community, the Nordic countries, Canada, the USA, and certain other countries have called for this complete phase-out, rather than a mere reduction in use of the fully halogenated chlorofluorocarbons. Adjustments of the Protocol were agreed by the Parties to the Protocol in June 1990. A total phase-out of 15 fully halogenated chlorofluorocarbons, halons 1211, 1301, and 2402, and carbon tetrachloride is to be effected by the year 2000. In addition, methyl chloroform must be phased out by the year 2005.

These developments have created an urgent need for acceptable substitute chemicals. These should have similar physical and chemical properties and safety characteristics to the chlorofluorocarbons included in the Montreal Protocol. There should be a realistic anticipation that their commercial-scale production will be technologically and economically feasible, and their ozone-depleting potential and possible global-warming potential should be considerably lower.

The phase-out of chlorofluorocarbons can also be accomplished by employing alternative technologies.

The chemical industry worldwide is already engaged in efforts to develop substitutes for the chlorofluorocarbons included in the Montreal Protocol. In order to avoid the risk of introducing chemicals that could prove to be either health or environmental hazards, the toxicological and environmental evaluation of the potential substitutes is of utmost importance and urgency. There are already two international industry-supported efforts underway: the Programme for Alternative Fluorocarbon Toxicity Testing (PAFT), and the Alternative Fluorocarbon Environmental Acceptability Studies (AFEAS).

There is a need to help prevent the use of harmful chlorofluorocarbons and of any substitutes for the harmful chlorofluorocarbons that would pose an unreasonable risk to human health or the environment. There is also a need for producers to make decisions about the manufacture of acceptable substitutes in time. Environmental Health Criteria 113: Fully Halogenated Chlorflurocarbons (WHO, 1990) evaluated ten fully halogenated chlorofluorocarbons. Among these are the five compounds included in the Montreal Protocol mainly on the basis of their high ozone-depleting potential and long residence times in the atmosphere. The ozone-depleting and global-warming potentials of the partially halogenated chlorofluorocarbons are considerably lower and their atmospheric residence times are shorter. Thus certain of these partially halogenated chlorofluorocarbons, i.e. those for which the toxicity evaluation suggests no unreasonable health or environmental risk and which are likely to be technologically and economically feasible, could be possible substitutes for the fully halogenated derivatives. In this monograph, the

Introduction

evaluation of two partially halogenated chlorofluorocarbons (methane derivatives) is presented. The evaluation of six other partially halogenated chlorofluorocarbons (ethane derivatives) has already started and will shortly be published in the Environmental Health Criteria series. The mere selection of a chemical for the evaluation programme of the IPCS does not mean its endorsement as a substitute chemical. Only a full toxicological and environmental evaluation can be a basis for such a conclusion.

1. SUMMARY

1.1 Identity, physical and chemical properties, and analytical methods

The two chlorofluorocarbons reviewed in this monograph (dichlorofluoromethane, HCFC 21, and chlorodifluoromethane, HCFC 22) are hydrochlorofluorocarbons (HCFCs), i.e. compounds derived by the partial substitution of the hydrogen atoms in methane with both fluorine and chlorine atoms. Only HCFC 22 has commercial significance. Both HCFC 21 and HCFC 22 are non-flammable gases (at normal temperatures and pressures), colourless, and practically odourless. HCFC 21 is slightly soluble and HCFC 22 moderately soluble in water, and both are miscible with organic solvents. HCFC 22 is available as a liquified gas.

There are several analytical methods for determining these two HCFCs. These include gas chromatography with electron capture and flame ionization detection, gas chromatography/mass spectrometry, and photothermal deflection spectrophotometry.

1.2 Sources of human and environmental exposure

The two HCFCs reviewed in this monograph are not known to occur as natural products. HCFC 21 is only produced in small quantities for non-occupational purposes. The total annual worldwide production of HCFC 22 was estimated in 1987 to be 246 000 tonnes.

The main loss of HCFC 22 is due to its release during the repair, use, and disposal of refrigerators and air-conditioning units. The estimated maximum current worldwide loss is around 120 000 tonnes per year. There have been reports of accidental release of HCFC 22 on fishing vessels.

HCFC 22 is used as a refrigerant, as an intermediate in the production of tetrafluoroethylene, and as a blowing agent for polystyrene. A small quantity is used as an aerosol propellant.

Summary

1.3 Environmental transport, distribution, and transformation

The log octanol/water partition coefficient for HCFC 22 is 1.08, which makes bioaccumulation unlikely. The estimated tropospheric lifetime of HCFC 21 is about 2 years and that of HCFC 22 about 17 years. Reaction with hydroxy radicals in the troposphere is likely to be the primary route of degradation. Only a small fraction of HCFCs 21 and 22 reach the stratosphere, where, mainly by reaction with oxygen radicals, they release ozone-depleting chlorine. However, it is estimated that HCFC 22 is responsible for less than 1% of the ozone-depleting chlorine in the stratosphere. The ozone-depleting potential (ODP) of HCFC 22 has been estimated to be 0.05, while that of HCFC 21 is assumed to be lower.

The global-warming potential (GWP), relative to CFC 11 (taken as 1.0), has been estimated to be lower by a factor of about 3-4 for HCFC 22 and lower still for HCFC 21.

1.4 Environmental levels and human exposure

There are no data available on concentrations in water or on the presence of these compounds in food, although HCFC 22 is used in the manufacture of expanded polystyrene food containers. There are no data on human exposure to HCFC 21, but two studies on the use of experimental sprays containing 17-65% HCFC 22 have shown that short (10-20 secs) exposures might result in peak concentrations ranging from 5000 to 8000 mg/m^3. Workers in beauty parlours could be exposed to 8-h time-weighted average levels of 90-125 mg/m^3, but these are well below the currently regulatory MAK or LTV levels of 1800-3540 mg/m^3 for Germany, USA, and the Netherlands.

HCFC 22 mixes rapidly in the atmosphere. Concentrations of about 326 mg/m^3 were reported in 1986 and the level is believed to be increasing by about 11% annually.

1.5 Kinetics and metabolism in laboratory animals and humans

There are limited data on the absorption, distribution, metabolism, and excretion of HCFC 21. That HCFC 21 is absorbed following inhalation can be inferred from systemic effects and the elevated urinary fluoride levels seen in toxicity studies in rats. HCFC 21 is exhaled by rats following intraperitoneal injection, and both kinetic data and evidence from fluoride excretion suggest that HCFC 21 is metabolized. However, the extent of metabolism is unknown and, apart from fluoride, the products have not been identified.

HCFC 22 is rapidly and well absorbed following inhalation in rat, rabbit, and humans and is distributed widely. High levels of HCFC 22 have been found in the blood, brain, heart, lung, liver, kidney, and visceral fat of rabbits dying during exposure and in postmortem samples of brain, lung, liver, and kidney from accidental victims of HCFC 22 exposure. Elimination is rapid, most of the HCFC being eliminated with a half-life of 1 min in the rabbit and 3 min in the rat. In humans, a limited amount of material is eliminated in three phases (half-lives of 3 min, 12 min, and 2.7 h).

Inhaled or intraperitoneally administered HCFC 22 is almost entirely exhaled unchanged in both rats and humans. There is good evidence that no significant metabolism occurs *in vivo* in rats or in rat liver preparations.

1.6 Effects on laboratory mammals and *in vitro* test systems

There are no satisfactory data on the acute oral toxicity of HCFC 21 or HCFC 22.

The principal effects of a single inhalation exposure to HCFC 21 or HCFC 22 are essentially similar in a variety of animal species. Both substances have low toxicity by this route. Effects seen are typical of those of chlorofluorocarbons, i.e. loss of coordination and narcosis. Cardiac arrhythmias and pulmonary effects may occur at high concentrations (106.7 g/m^3 or more).

It has been claimed that both HCFC 21 and HCFC 22 cause skin and eye irritation, although these effects may have been related to the consequences of heat loss due to evaporation rather than to the chemical properties of the HCFCs. Neither substance caused skin sensitization.

The only studies conducted on the short-term toxicity of HCFC 21 have investigated the inhalation route. Liver

Summary

damage was the principal effect noted in the rat, guinea-pig, dog, and cat; a no-observed-effect level was not determined. Histopathological lesions of the liver were seen in rats at levels as low as 0.213 g/m^3 given 6 h/day, 5 days/week, for 90 days. Pancreatic interstitial oedema and seminiferous tubule epithelial degeneration also occurred at this level. Lesions were essentially absent in studies with HCFC 22 at exposure levels between 17.5 g/m^3 (for 13 weeks) and 175 g/m^3 (for 4 or 8 weeks).

There have been no long-term studies on HCFC 21 in animals. The only consistent non-tumorigenic finding in long-term studies with HCFC 22 was hyperactivity seen in male mice given 175 g/m^3, 5 h/day, 5 days/week in a lifetime inhalation study.

No conventional studies have investigated the effects of HCFC 21 on fertility. In an embryotoxicity study on rats (42.7 g/m^3, 6 h/day on days 6-15 of gestation) no teratogenic effect was observed, but a high rate of implantation loss was found. HCFC 22 (175 g/m^3 per day, 5 h/day, 5 days/week for 8 weeks) had no effect on the reproductive capacity of male rats. As a consequence of a small, non-significant excess of eye defects seen in three teratology studies in rats, an extensive study was conducted on the potential ability of HCFC 22 to cause eye defects. In this study a small, but statistically significant, increase in the number of litters containing fetuses with microphthalmia or anophthalmia was found following maternal exposure to 175 g/m^3, 6 h/day on days 6-15 of gestation. This exposure level gave slight maternal toxicity (lower body weight compared to controls). No other effects were seen, and 3.5 g/m^3 was the no-observed-effect level in this study. HCFC 22 was not teratogenic in a conventional study on rabbits at similar exposure regimens.

HCFC 21 was found to be non-mutagenic in two bacterial and one yeast assay (no further data was available). HCFC 22 was mutagenic in bacterial assays using *S. typhimurium*, but did not show activity in tests on other microorganisms or in mammalian systems, either *in vitro* or *in vivo*. These tests covered gene mutation and unscheduled DNA synthesis *in vitro, in vivo* bone marrow cytogenetic assays, and dominant lethal assays in both rat and mouse.

Carcinogenicity assays *in vivo* have only been conducted with HCFC 22. Two groups of investigators have conducted lifetime inhalation studies on both rats and mice. The only evidence of excess tumours occurred in the one study in which male rats were given 175 g/m^3, 5 days per week, for up to 131 weeks. Small excesses of fibrosarcomas of the salivary gland region and of Zymbal's gland were noted. These effects were not seen at lower doses (up to 35 g/m^3), and this high dose was not used in the second study. Although it was not an adequate demonstration of the absence of tumorigenic effects, no excess of tumours was seen in an oral gavage study on rats. These animals were given HCFC 22 at a level of 300 mg/kg per day, 5 days/week, for 52 weeks, and the study terminated at 125 weeks.

1.7 Effects on humans

Only very limited data are available on the effects of HCFC 21 and HCFC 22 in humans.

Death has occurred following accidental or intentional exposure to high levels of HCFC 22. Histopathological examination of the tissues of some of these victims revealed oedematous lungs and cytoplasmic fatty droplets mainly in the peripheral liver hepatocytes.

Although an increase in the incidence of palpitations has been claimed in a questionnaire study on people occupationally exposed to HCFC 22, there is no good evidence that volunteer or occupational exposure to HCFC 21 or HCFC 22 leads to ill health effects. No conclusions can be drawn from a very small mortality study on people occupationally exposed to several chlorofluorocarbons including HCFC 22.

1.8 Effects on other organisms in the laboratory and field

There are no data available on the effects of HCFCs 21 and 22 on organisms in the environment.

1.9 Evaluation and conclusions

Environmental exposure levels of both HCFC 21 and HCFC 22 are extremely low and are not considered likely to cause direct effects on human health. Controlled occupational exposures are also unlikely to represent a significant risk to humans.

Both HCFC 21 and HCFC 22 have a lower ozone-depleting potential and a shorter atmospheric residence time than the fully halogenated chlorofluorocarbons and should therefore pose a lower indirect health risk. Their global-warming potentials are considerably lower than those of the fully halogenated chlorofluorocarbons suggesting a lower environmental effect.

Since the toxicity of HCFC 22 is low, the ozone-depleting and global-warming potentials lower, and the atmospheric residence time shorter than those of the fully halogenated chlorofluorocarbons, it can be considered as a transient substitute for the CFCs included in the Montreal Protocol. Although HCFC 21 poses a low environmental and indirect health risk, it is not recommended as a substitute for the chlorofluorocarbons included in the Montreal Protocol because of possible direct health risk due to its liver toxicity.

2. IDENTITY, PHYSICAL AND CHEMICAL PROPERTIES, AND ANALYTICAL METHODS

2.1 Identity

The chlorofluorocarbons reviewed in this monograph are dichlorofluoromethane (HCFC 21) and chlorodifluoromethane (HCFC 22). These are hydrochlorofluorocarbons (HCFCs), i.e. compounds derived by the partial substitution of the hydrogen atoms in methane with both fluorine and chlorine atoms. The chemical formulae, chemical structures, common names, common synonyms, trade names and CAS registry numbers are presented in Table 1.

2.2 Physical and chemical properties

The physical and chemical properties of HCFCs 21 and 22 are summarized in Table 2. They are non-flammable gases at normal temperatures and pressures, colourless, and practically odourless. They are slightly or moderately soluble in water and miscible with organic solvents (Horrath, 1982; Weast, 1985).

Generally, hydrochlorofluorocarbons of low relative molecular mass are characterized by high vapour pressure, density, and refractive index, and low viscosity and surface tension (Bower, 1973).

HCFC 22 is available as a liquified gas with a minimum purity of 99.9% or in a variety of blends and azeotropic mixtures.

2.3 Conversion factors

Conversion factors for HCFCs 21 and 22 are given in Table 1.

2.4 Analytical methods

Analytical procedures for the determination of HCFCs 21 and 22 are summarized in Table 3. By far the most frequently applied methods use gas chromatography with various detection techniques.

Table 1. Identity of HCFC 21 and HCFC 22[a]

	HCFC 21	HCFC 22
Chemical structure	H−C(Cl)(Cl)−F	H−C(Cl)(F)−F
Chemical formula	$CHCl_2F$	$CHClF_2$
Common name	dichlorofluoromethane	chlorodifluoromethane
Common synonyms and trade names	methane, dichlorofluoro-; fluorodichloromethane; F-21; R-21; Freon 21; Genetron 21; dichloromonofluoromethane; monofluorodichloromethane	Algeon 22; Arcton 22; Chlorofluorocarbon 22; difluorochloromethane; difluoromonochloromethane; Electro-CF 22; Eskimon 22; F-22; FC-22; Flugene 22; Fluorocarbon 22; Forane 22; Freon 22; Frigen 22; Genetron 22; HFA 22; Hydrochlorofluorocarbon 22; Hydrofluoroalkane 22; Isceon 22; Osotron 22; Khladon 22; methane, chlorodifluoro-; monochlorodifluoromethane; Propellant 22; R-22; Refrigerant 22; UCON 22
CAS registry number	75-43-4	75-45-6
Conversion factors (20 °C)		
ppm → mg/m^3	4.276	3.54
mg/m^3 → ppm	0.234	0.282

[a] Chlorofluorocarbons are numbered as follows: the first digit = number of C atoms minus 1 (for methane derivatives it is therefore zero); second digit = number of H atoms plus 1; third digit = number of F atoms.

A number of methods have been described for the determination of HCFC 21. These include gas chromatography with dual flame detection (Lindberg, 1979) and with electron capture detection (Vidal-Madjar et al., 1981; Rasmussen et al., 1983 and Höfler et al., 1986). Similarly, a number of methods have been described for the determination of HCFC 22. These include gas chromatography/mass spectrometry (Brunner et al., 1981), gas chromatography with electron capture detection (Shimohara et al., 1979), and photothermal deflection spectrophotometry (Long & Bialkowski, 1985).

Identity, Physical and Chemical Properties, and Analytical Methods

Table 2. Physical and chemical properties of HCFC 21 and HCFC 22[a]

	HCFC 21	HCFC 22
Physical state	gas	gas
Colour	colourless	colourless
Relative molecular mass	102.92	86.47
Boiling point (°C) at 103 kPa	8.9	- 40.8
Freezing point (°C)	-135.0	-146.0
Liquid density (g/ml)	1.405 (at 9 °C)	1.49 (at -68 °C)
Vapour density (g/litre) at boiling point	4.57	4.82
Vapour pressure (atm) at 21 °C	1.57	9.33
Surface tension (dynes/cm) at -41 °C	-	15
Refractive index at 9 °C	1.3724	-

[a] From: Grasselli & Richey (1975); Hawley (1981); Horrath (1982); Sax (1984); Weast (1985).

Table 3. Analytical methods for the determination of partially halogenated methane derivatives

Medium	Analytical method	Detection limit	Reference
HCFC 21			
Air	gas chromatography with dual flame ionization detection		Lindberg (1979)
	gas chromatography with flame ionization detection		NIOSH (1985)
	gas chromatography with electron capture detection		Höfler et al. (1986)
	gas chromatography with electron capture detection	0.009 $\mu g/m^3$	Rasmussen et al. (1983)
	gas chromatography with electron capture detection	0.08 $\mu g/m^3$	Vidal-Madjar et al. (1981)
HCFC 22			
Air	gas chromatography with electron capture detection	0.14-0.46 $\mu g/m^3$	Shimohara et al. (1979)
	gas chromatography/ mass spectrometry	0.4 $\mu g/m^3$	Brunner et al. (1981)
	photothermal deflection spectrophotometry	0.6 $\mu g/m^3$	Long & Bialkowski (1985)
Blood and tissues	head space method, gas chromatography with flame ionization detection	0.1 $\mu l/g$	Sakata et al. (1981)
	head space method, gas chromatography with flame ionization detection		Morita et al. (1977)

3. SOURCES OF HUMAN AND ENVIRONMENTAL EXPOSURE

3.1 Natural occurrence

HCFC 21 and HCFC 22 are not known to occur as natural products.

Stoibe et al. (1971) reported the presence of HCFC 22 in volcanic emissions, but Rasmussen et al. (1980) did not observe any excess of this compound, compared with normal atmospheric levels, in their studies of volcanic emissions. In their analyses of air samples collected over the State of Washington, USA, Leifer et al. (1981) found that the concentrations of HCFC 22 (110-195 ng/m^3) after the eruption of the Mount St. Helens volcano were no higher than normal.

3.2 Anthropogenic sources

3.2.1 Production levels

HCFC 21 has been manufactured by one company in the USA, but only in very small quantities, and is no longer produced (personal communication by H. Trochimowicz to IPCS, 1990).

HCFC 22 is produced by companies in the USA, Western Europe, Japan, Latin America, and elsewhere, the total annual world-wide production in 1987 being estimated to be 246 000 tonnes (personal communication by E.I. Du Pont de Nemours & Co. Inc., 1989).

3.2.2 Manufacturing processes

Both HCFC 21 and HCFC 22 are manufactured by the liquid phase reaction of chloroform with anhydrous hydrofluoric acid in the presence of an antimony halide catalyst (Hawley, 1981) at various reaction temperatures and pressures (SRI, 1985). This process is being replaced by a continuous vapour-phase process employing gaseous hydrogen fluoride in the presence of chromium oxide or halide, ferric chloride, or thorium tetrafluoride catalysts (Grayson, 1978).

3.2.3 Loss during disposal, transport, storage, and accidents

The source of the main loss of HCFC 22 during disposal is discarded refrigerators and air conditioners. Trace quantities of HCFCs 21 and 22 have been detected in landfill gas (Höfler et al., 1986).

Equipment for the transport and storage of HCFC 22 is designed to withstand high pressure and is fitted with safety valves, bursting discs, and fusible plugs. Losses of product during normal transport and storage should, therefore, be relatively small because of the completely closed systems used.

The main release of HCFC 22 occurs in the form of leakages from refrigeration and air-conditioning units (see section 3.3.2). Two accidental releases with fatal consequences on fishing vessels have been described by Morita et al. (1977) and Haba & Yamamoto (1985). In general, some fugitive losses during manufacturing are likely.

3.3 Use patterns

3.3.1 Major uses

HCFC 21 has been reported to be used as a refrigerant for centrifugal machines, as a solvent (where its high Kauri-butanol number is desirable), in combinations with Freon 12 in aerosol products (National Library of Medicine, 1990), in fire extinguishers (Hawley, 1981), as a propellant gas (Sittig, 1985), and as a heat exchange fluid in geothermal energy applications (Grayson, 1978). However, it is no longer manufactured for any commercial purpose (personal communication by H. Trochimowicz, 1990).

HCFC 22 is used as a refrigerant in residential, commercial, and mobile air-conditioning units. An azeotropic mixture (HCFC 502) of HCFC 22 and CFC 115 (48.8:51.2 wt.%) is used as a refrigerant in food display cases, ice makers, home freezers, and heat pumps (American Chemical Society, 1985). It is also used as an intermediate in the production of tetrafluoroethylene by pyrolysis at 650-700 °C (Smart, 1980). It is estimated that approximately 34% of the total amount is currently used for this latter purpose. HCFC 22 is used as a blowing agent, especially for polystyrene (see section 5.1.3) and as a propellant in aerosols (Hanhoff-Stemping, 1989). It is not used to any significant extent as an industrial solvent but has been in the past (US EPA, 1981).

3.3.2 Releases during use: controlled or uncontrolled

Data are only available for HCFC 22. The major loss to the environment results from equipment and system leaks during use, repair, servicing, and after scrapping (Salzburger et al., 1989). Assuming no significant loss from its use as a polymer intermediate, it has been estimated that the maximum current worldwide release of HCFC 22 is around 120 000 tonnes per year (personal communication by E.I. Du Pont de Nemours & Co. Inc., 1988).

4. ENVIRONMENTAL TRANSPORT, DISTRIBUTION, AND TRANSFORMATION

Environmental Health Criteria 113: Fully Halogenated Chlorofluorocarbons considered the transport between media, environmental transformation processes, interaction with other physical, chemical or biological factors, and bioconcentration and bioaccumulation of fully halogenated chlorofluorocarbons (WHO, 1990). Much more information has been published on fully halogenated than on partially halogenated chlorofluorocarbons.

4.1 Biodegradation and bioaccumulation

There is practically no information on the biodegradation in the environment of HCFC 21 and HCFC 22. The log octanol/water partition coefficient of HCFC 22 is 1.08, which makes bioaccumulation of this hydrochlorofluorocarbon unlikely (Hansch & Leo, 1979).

4.2 Environmental transformation and interaction with other environmental factors

The physical and chemical properties of the partially halogenated chlorofluorocarbons suggest that they would mix rapidly within the lower region of the troposphere. Mixing would be expected to be complete in the hemisphere of the emission (northern or southern) within months and in the entire troposphere possibly within about three years. The tropospheric concentration of HCFC 22 is rapidly increasing (Hanhoff-Stemping, 1989). Reaction with naturally occurring hydroxy radicals in the troposphere is thought to be the primary degradation route. The estimated tropospheric rate of this reaction is such that the average lifetime is about 2 years for HCFC 21 (UNEP/WMO, 1989) and 13-25 years for HCFC 22 (Makide & Rowland, 1981; WMO, 1986; UNEP/WMO, 1989; Zurer, 1989). It should be noted that the lifetimes of fully halogenated chlorofluorocarbons are much longer (75 years for CFC 11 and 111 years for CFC 12). The mechanism of decomposition of HCFC 22 following the initial reaction with hydroxy radicals has been studied but not fully elucidated. The most likely product of the gas-phase reaction in the atmosphere is carbonyl fluoride (COF_2) (Atkinson, 1985), which would be hydrolysed rapidly by atmospheric water to carbon dioxide and hydrogen fluoride, the latter being removed by precipitation.

The small fraction of HCFC 22 not destroyed in the troposphere slowly enters and mixes with the upper layer of the atmosphere, the stratosphere. Seigneur et al. (1977), in their discussion of the photochemical and chemical processes of HCFCs 21 and 22 in the atmosphere, used a one-dimensional model at steady state to estimate the upward diffusion of these HCFCs in the atmosphere from ground level to 60 km. Their models of diffusion and reaction predict that, at steady state, the amounts of these chemicals reaching the stratosphere would be, relative to the amount released at ground level, between 1 and 3% for HCFC 21 and between 4 and 12% for HCFC 22. The major destruction mechanisms are reactions with hydroxy radicals and excited oxygen atoms (Seigneur et al., 1977). Photo-decomposition by solar ultraviolet radiation, which is a major process for the fully halogenated chlorofluorocarbons, does not play a significant role in the destruction of HCFC 22 in the stratosphere (Molina et al., 1976).

The major part of the ozone-depleting chlorine in the stratosphere comes from fully halogenated chlorofluorocarbons. The above model indicates that the partially halogenated chlorofluorocarbons are not expected to have high ozone-depleting potential (ODP). The ODP is defined as the calculated ozone depletion due to the emission of a unit mass of the chlorofluorocarbon divided by the ozone depletion calculated to be due to the emission of CFC 11; calculations are based on steady-state conditions (UNEP/WMO, 1989).

No ODP value has been determined for HCFC 21 although Seigneur et al. (1977) stated that the compound is "50-100 times less hazardous than CFC 11" to the stratospheric ozone level, based on calculations by Molina et al. (1976). The ODP for HCFC 22 has been estimated to be 0.05 (Hammitt et al., 1987; UNEP, 1988). Solomon & Tuck (1990) believe that reaction on ice particles in the Antarctic stratosphere may raise this figure by a factor of two or more, but Fisher et al. (1990b) do not completely agree with this supposition. A value of 0.05 means that continuous emissions of HCFC 22 would have to be 20 times as large as continuous emissions of CFC 11 to have the same effect on ozone. A 5% annual increase in HCFC 22 emissions has been estimated by Ramanathan et al. (1985), but the 11-12% figure calculated by Khalil & Rasmussen (1981, 1983) agrees better with the published data. The 16% yearly increase assumed by Krüger & Fabian (1986) is probably too high. The transitional use of HCFC 22 as a substitute for the highly ozone-depleting fully halogenated CFC 12 has been agreed, but the phase-out of HCFC 22 is also foreseen (Zurer, 1990; UNEP, 1990; FRG, 1990).

Current assessment of the global-warming potential (GWP) ("greenhouse effect") of HCFC 22, based on the comparison with CFC 11 (the reference compound with a GWP of 1.0), indicates that it is lower by a factor of about 3-4 (Garber, 1989; Fisher et al., 1990a). The GWP of HCFC 21 is lower still (Garber, 1989).

5. ENVIRONMENTAL LEVELS AND HUMAN EXPOSURE

5.1 Environmental levels

5.1.1 Air

Penkett et al. (1980) reported that extremely low background concentrations of HCFC 21, ranging from 4.3 to 8.6 ng/m^3 (1-2 ppt), have been found in the atmosphere. However, other investigators have determined much higher concentrations ranging from 43 to 86 ng/m^3 (10-20 ppt) (Crescentini & Bruner, 1979). Rasmussen et al. (1983) concluded that this difference in the reported concentrations of HCFC 21 in the troposphere cannot be reconciled on the basis of difficulties in identification or differences in absolute accuracy.

Rasmussen et al. (1980) determined an average global HCFC 22 concentration of 159 ng/m^3 (45 ppt) in mid-1979. The average concentrations were 177 ng/m^3 (50 ppt) and 149 ng/m^3 (42 ppt) in the northern and southern hemispheres, respectively. These values are considerably higher than the value of 89-106 ng/m^3 (25-30 ppt) calculated from estimates of emissions. Leifer et al. (1981) found values (110-190 ng/m^3) over the State of Washington, USA, in 1980.

Khalil & Rasmussen (1981) determined concentrations of HCFC 22 in 100 atmospheric samples collected between April 1978 and January 1981 at a latitude of 45 °N in the northwest Pacific. The concentration of HCFC 22 increased at an average rate of 11.7% per year over the two and a half years of the study, and in January 1981 was about 230 ng/m^3 (65 ppt). The same authors calculated the concentrations expected from the estimated industrial release of HCFC 22 since 1950. The observed concentrations were, on average, 60 ng/m^3 (17 ppt) higher than the estimated concentrations. The authors considered the difference to be the consequence of an underestimation of past industrial release.

Rasmussen & Khalil (1983) measured an average HCFC 22 concentration of 259 ng/m^3 (73 ppt) in the lower Arctic atmosphere (0-4 km) at 70 °N in May 1982. This concentration was 1.3 times greater than that found at 30-40 °S in November 1981 (Rasmussen et al., 1982), a difference which the authors considered significant.

In the Arctic (72 °N in Alaska), the winter concentrations of HCFC 22, as well as those of other halocarbons, carbon monoxide, and soot from combustion, are higher than at other times of the year. This is attributable to faster transport of anthropogenic emissions in winter. The average winter concentration of HCFC 22 in the years 1980 and 1981 was 217 ng/m^3 (61 ppt), whereas the average during the summer was 198 ng/m^3 (56 ppt). The rate of increase from August 1980 to February 1982 was 11.9% per year (Khalil & Rasmussen, 1983). The most recent studies reported a concentration of 326 ng/m^3 (92 ppt) in 1986 (NASA, 1988), which corresponds approximately to the expected value based on an average annual increase of 11% since 1979. The rate of increase is somewhat uncertain due to the limited number of measurements, and lower values have been assumed by other authors (Ramanathan et al., 1985). These authors pointed out that, at an estimated increase of 5% per year, the average concentration in the global atmosphere would reach 3200 ng/m^3 by the year 2030.

5.1.2 Water

No data are available on the concentrations in water of the HCFCs 21 and 22.

5.1.3 Food and other edible products

No information is available on the possible content of partially halogenated chlorofluorocarbons in food or other edible products.

HCFC 22 is used in the USA, the United Kingdom, and several other countries as a blowing agent for polystyrene foam, an authorized food contact plastic.

5.2 General population exposure

There are no data on human exposure to HCFC 21. The maximum workplace concentration (MAK) is limited to 45 mg/m^3 in Germany (DFG, 1990) and to 42 mg/m^3 in the USA (ACGIH, 1990).

Simulated-use studies have been carried out to assess the potential human exposure to HCFC 22 arising from its assumed use as an aerosol propellant. After a single spray (5 or 10 seconds duration) of an aerosol containing 17% HCFC in a closed room of 22 m^3, the air concentration was determined at various positions relative to the spray cone. When the spray was directed towards the sampling tube, peak concentrations were 5075 and 8050 mg/m^3 after 5 and 10 seconds spraying, respectively. The concentrations declined after about 10 or 20 seconds, respectively, stabilizing at levels of 25 or 45 mg/m^3. In all other spray positions the concentration did not exceed the level calculated for homogeneous distribution in the air of the room (Bouraly & Lemoine, 1988).

Hartop & Adams (1989) reported a series of similar studies in which the concentrations of HCFC 22 were measured during simulated human use using experimental manikins representing adult and child. They examined hair sprays containing 20-40% HCFC 22, whole body deodorants containing 20-65%, and antiperspirants containing 20-40%. The peak concentrations found in a closed room of 21 m^3 ranged from 53 mg/m^3 (for a 4-second spray of an antiperspirant containing 18.8% HCFC 22) to 5000 mg/m^3 (for a 20-second spray of a deodorant containing 65%). This corresponds to 10-min weighted average concentrations of about 50 mg/m^3 and 1440 mg/m^3, respectively. In the same study, simulated use of a hair spray in a beauty parlour gave 10-min weighted average values of 160-225 mg/m^3 for the "customer" and 8-h weighted average values of 90-125 mg/m^3 for the "beautician", based on the assumption that the latter would use one 10-second spray every 15 min with the door of the beauty parlour open. This latter value is well below the MAK (Maximale Arbeitsplatzkonzentration, maximum working place concentration) of 1800 mg/m^3 in Germany (DFG, 1990) and the TLV of 3540 mg/m^3 in the USA (ACGIH, 1990).

5.3 Occupational exposure

Bales (1978) reported HCFC 22 exposure levels of 17-48 mg/m^3 for workers in a fluorocarbon packaging and shipping plant.

6. KINETICS AND METABOLISM IN LABORATORY ANIMALS AND HUMANS

6.1 Animal studies

6.1.1 Absorption

There are no quantitative data on the absorption of HCFC 21, but the increase in fluoride levels observed in 90-day inhalation toxicity studies (section 7.2) and the data on metabolic transformation (section 6.1.3) suggest that inhaled HCFC 21 is absorbed.

Carney (1977) studied the relationship between exposure to HCFC 22 and blood levels in anaesthetized rats. The gas in air was applied through a canula inserted into the trachea. After 15 min of exposure, a blood sample was withdrawn from the carotid artery, and further blood samples were taken at intervals up to 30 min. Male and female rats were exposed to nominal air concentrations of 35 or 175 g/m^3. At 35 g/m^3 the mean blood concentration was 31 mg/litre, while at 175 g/m^3 it was 155 mg/litre, showing a relationship between the inhaled air and blood concentrations of HCFC 22. The clearance was rapid with a half-life of 3 min.

Similar results have been reported by Sakata et al. (1981) in experiments with rabbits. Animals, anaesthetized with phenobarbitone (25 mg/kg ip), received HCFC 22/air mixtures via a plastic mask, and blood samples were taken through a catheter in a femoral artery. The concentration of HCFC 22 inhaled ranged from 175 g/m^3 (5%) to 1400 g/m^3 (40%). At every air concentration, blood concentrations increased rapidly from the beginning of inhalation, and saturation was reached in about 5 min. The blood concentration was related to the concentration in inhaled air: at 175 g/m^3 it was 148 mg/litre (similar to the blood level found by Carney (1977) in rats under the same conditions), and at 700 g/m^3 it was 583 mg/litre. Clearance was rapid with a half-life of 1 min; no HCFC 22 could be detected one hour following cessation of exposure.

Woollen (1988) exposed pregnant rats to atmospheric HCFC 22 concentrations of between 1239 and 609 g/m^3 (350 and 175 000 ppm). Blood samples taken at various intervals showed that the compound rapidly reached equilibrium with the blood. At the highest exposure level, the blood level reached 118.5 mg/litre after 30 min and there was no significant increase after a further 5.5 h of exposure (121 mg/litre).

6.1.2 Distribution

No information is available on the distribution of HCFC 21.

Sakata et al. (1981) determined the amount of HCFC 22 in the tissues of rabbits exposed by inhalation to concentrations of up to 1400 g/m^3 (see section 6.1.1). The results are presented in Table 4. No major differences were found in the tissues examined except for fat tissue, where there was a difference after long and short inhalation times at high concentrations. The authors postulated that the effect was related to the poor vascular blood supply of adipose tissue and that the distribution depended on partial pressures. Those tissues with a good blood supply would reach equilibrium quickly, whereas fat would equilibrate only slowly. Komoriya et al. (1980) found HCFC 22 to be widely distributed in rats after a variety of lethal exposures.

6.1.3 Metabolic transformation

A saturable dose-related increase in urinary fluoride was observed in both sexes of Charles River albino rats in a 90-day inhalation study with HCFC 21 (Lindberg, 1979). Details are given in section 7.2.

The pharmacokinetics of HCFC 21 has been investigated in male Wistar rats. Animals were injected intraperitoneally with a dose of 3.25 ml gas/kg body weight and placed in a closed chamber. Exhaled HCFC 21 was monitored for 7 h by gas chromatography. Only part of the injected compound was exhaled and it was assumed that the remainder was metabolized (Peter et al., 1986).

Peter et al. (1986) found that HCFC 22 was not metabolized by Wistar rats after intraperitoneal exposure (3.08 ml gas/kg body weight). In their first experiment,

Table 4. Tissue levels of HCFC 22 in rabbits following lethal inhalation exposure[a]

	Rabbit no.							
	1	2	3	4	5	6	7	8
HCFC 22 concentration (%, v/v)[c]	0-32	0-29	0-33	0-42	30	30	40	40
O_2 concentration (%, v/v)	20-14	20-14	20-13	20-14	20	20	20	20
Time of death after start of inhalation (min)	67	31	25	19	15	92	7	10
Amount of HCFC 22[b]								
Brain	145	138	76	156	137	148	140	159
Heart	145	150	100	158	135	140	125	129
Lung	167	128	187	231	139	136	121	186
Liver	143	95	72	78	101	153	44	60
Kidney	160	90	81	89	102	142	82	56
Visceral fat	327	93	48	33	38	196	23	27
Blood	219	193	140	161	131	219	147	199

[a] From: Sakata et al. (1981).
[b] Values are µl/g at 20-25 °C, 1 atmosphere
[c] Animals 1-4 were exposed to increasing concentrations of HCFC 22. Values are inhaled concentrations; highest concentrations in rabbits 1-4 indicate those at the time of death.

rats received a single ip injection of HCFC 22, after which they were placed in a closed desiccator with a gas sample loop connected to a gas chromatograph. The injected HCFC 22 was almost completely exhaled. Pretreatment of the animals with phenobarbital (80 mg/kg ip followed by 3 days with 0.1% phenobarbital in the drinking-water) or DDT (200 mg/kg, one week prior to the experiment) did not alter the observation. The authors concluded that there was no detectable metabolism of HCFC 22. From these studies, it seems that HCFC 22, unlike HCFC 21, is not metabolized.

The findings of Peter et al. (1986) support the earlier, more comprehensive, studies of Salmon et al. (1979) who carried out *in vivo* studies using ^{14}C- and ^{36}Cl-labelled HCFC 22. Alderley Park Wistar-derived rats were exposed to ^{14}C-labelled HCFC 22 at levels of 1.75

g/m^3 in three experiments and 35 g/m^3 in three others. The exposure durations were 15-24 h. Exhaled carbon dioxide was collected by absorption on barium hydroxide and radioactivity was subsequently measured. Separate collection of urine and faeces into containers cooled to 0 °C was followed by measuring radioactivity, directly in the case of urine and after appropriate oxidation in the case of faeces. Similar exposure and collection conditions were used for the experiments with ^{36}Cl-labelled HCFC 22, in which the exposure applied was 35 g/m^3 for 17.5 h. The study showed that metabolism of HCFC 22 in the rat was minimal. The amount of $^{14}CO_2$ released was equivalent to approximately 0.1% of the inhaled HCFC 22 at the exposure level of 1.75 g/m^3 and 0.06% at 35 g/m^3. The amounts of ^{14}C in the urine were also small, equivalent to approximately 0.03 and 0.01% of the inhaled dose at 1.750 and 35 g/m^3, respectively. Insignificant quantities were found in the faeces. The results of the experiment with ^{36}Cl supported those obtained with ^{14}C; only 0.01% of the inhaled dose was detected in urine. It is not quite clear whether the minimal metabolism observed was of HCFC 22 or of an impurity present in the test compound (Salmon et al., 1979).

Salmon et al. (1979) also conducted *in vitro* studies, using a microsomal preparation derived from liver homogenates of the same rat strain induced with Aroclor 1254. Microsomes, NADPH, and ^{36}Cl-labelled HCFC 22 were incubated in a repeat-dosing syringe, and samples were taken for analysis at 2-min intervals. Released ^{36}Cl$^-$ was isolated as silver chloride and estimated by scintillation counting. Under the test conditions, there was no release of chloride ion from HCFC 22 (studied over the concentration range 0.2-1.3 mmol/litre). This was a further indication of the resistance of HCFC 22 to breakdown in biological systems and suggested that any potential biological activity of the compound was unlikely to be due to the formation of reactive intermediates.

6.1.4 Elimination

Peter et al. (1986), in their pharmacokinetic studies of HCFC 21 and HCFC 22 (details in section 6.1.3), calculated the total clearance values of these compounds to be 4400 and 120 ml/h per kg, respectively.

Carney (1977) showed that the clearance of HCFC 22 from the blood of rats was rapid, the half-life being approximately 3 min (details given in section 6.1.1). In studies by Sakata et al. (1981) on rabbits, using exposure concentrations of 175 and 1400 g/m³, it was found that the blood concentration decreased rapidly after cessation of exposure, the maximum half-life being 1 min. After 15-30 min, blood concentrations were 27-31 mg/litre irrespective of the concentration inhaled. Once the exposure ceased, HCFC 22 was rapidly cleared from the blood and alveolar air, this being followed by slower elimination from poorly perfused tissues.

Studies of Salmon et al. (1979) demonstrated that only minimal amounts of the dose were excreted in the urine of rats following exposure to a HCFC 22 concentration of 35 g/m³. Peter et al. (1986) demonstrated that after ip injection the compound was exhaled unchanged almost completely (see section 6.1.3).

6.2 Human studies

Data are available on the absorption and elimination of HCFC 22 only.

6.2.1 Absorption and elimination

In studies by Woollen et al. (1989), two groups of three male subjects were exposed to average air HCFC 22 concentrations of 0.32 or 1.81 g/m³ for 4 h. Blood and expired air samples were collected during the exposure period and for up to 26 h after exposure, and were analysed for HCFC 22. Urine samples were collected for up to 22 h after exposure and analysed for HCFC 22 and fluorides. During the exposure period blood concentrations approached a plateau, the maximum blood concentrations of 0.25 and 1.36 µg/ml being related to the exposure level. The concentrations of HCFC 22 in the expired air were similar to the air concentrations during the exposure period. The ratio between blood and expired air concentrations towards the end of the exposure period was, on average, 0.77. This is consistent with *in vitro* measurements of the solubility of HCFC 22 in human blood (blood/air partition coefficient: 0.79). In the post-exposure period, three phases of elimination were apparent with half-lifes of 3 min, 12 min, and 2.7 h. The first phase, identified only from expired air analyses, probably represented elimination from alveolar air and/or lung tissues. The second and third phases may correspond to elimination from better and more poorly perfused tissues, respectively. HCFC 22 was detected in urine samples collected in the post-exposure period at both exposure levels, and the rate of decline was consistent with the terminal rate of elimination determined by blood and breath analyses. Fluoride concentrations in urine did not increase significantly following exposure, indicating that no detectable HCFC 22 metabolism occurs at these exposure levels.

6.2.2 Distribution

Three days after a fatal accident on board a fishing vessel (for details see section 8.2), samples of major tissues taken from two of the deceased people were analysed for HCFC 22 by gas chromatography (Morita et al., 1977). The findings are presented in Table 5. The concentrations were similar to those found in two rabbits examined three days after death by asphyxiation with HCFC 22 (Sakata et al., 1981; see section 6.1.2, Table 4).

Table 5. Concentration (μg/g) of HCFC 22 in major human tissues after fatal poisoning[a]

	Brain	Lung	Liver	Kidney	Blood
Subject A	68	18	71	18	69
Subject B	100	20	92	8	130

[a] From: Morita et al. (1977).

In a survey of organic compounds in human milk, HCFC 22 was detected in one of twelve samples as one of 184 compounds (Pellizzari et al., 1982). No information on exposure or quantification of the amount found was given in the report.

7. EFFECTS ON LABORATORY MAMMALS AND IN VITRO TEST SYSTEMS

7.1 Single exposure

7.1.1 Acute oral toxicity

No published data on the oral toxicity of HCFC 21 or HCFC 22 are available except for a study by Anatova et al. (1983), in which no signs of toxicity were noted in rats administered 4 ml of an aqueous HCFC 22 solution at a concentration of 2700 mg/litre.

7.1.2 Acute inhalation toxicity

Detailed information on the acute effects of HCFC 21 in various animal species is given in Table 6.

The signs of acute intoxication indicate that the CNS is the major target organ when animals are exposed to high concentrations of HCFC 21. Levels higher than 400 g per m^3 were lethal to the rat and guinea-pig within a few minutes to two hours. The 4-h LC_{50} in rats was 213.07 g per m^3 (Tappan & Waritz, 1964). Animals exposed to concentrations above 42.7 g/m^3 for 5 min or more exhibited signs typical of various stages of anaesthesia. Dyspnoea was observed at exposure levels above 50 g/m^3.

In addition to central nervous system depression, increased lacrimation, piloerection, and mydriasis were observed.

The concentrations and durations of exposure at which HCFC 22 proved lethal to a variety of animal species are given in Table 7. These data show that HCFC 22 has a low order of acute toxicity to several laboratory animal species. Deaths have been reported in rats, mice, and guinea-pigs exposed to HCFC 22 concentrations of 775-1295 g/m^3 (220 000 to 365 000 ppm) of HCFC 22 for periods of 15-240 min (Table 7). The signs of toxicity in rats were tremor of the limbs and head, convulsions, narcosis, shallow respiration, and death from respiratory depression. Death always occurred during exposure, never after. Recovery from non-lethal exposure was rapid. Rats appeared normal within 10 min and showed no delayed after-effects.

Table 6. Acute effects of HCFC 21 in various animal species

Species	Concentration (g/m³)	Time of exposure	Symptoms	Reference
Mouse	42.7	30-100 min	hyperactivity	Booth & Bixby (1932)
Rat	213.07	4 h	(LC_{50}); central nervous system depression, lacrimation, piloerection, mydriasis	Tappan & Waritz (1964)
Rat and guinea-pig	427	15-50 min	deep narcosis, death	Weigand (1971)
	213.5	2 h	loss of balance, narcosis	Weigand (1971)
	106.75	2 h	loss of balance, tremors, excitation	Weigand (1971)
	10	2 h	no changes	Weigand (1971)
Guinea-pig	1708	6 min	tremors, death	Booth & Bixby (1932)
	854	11 min	tremors, death	Booth & Bixby (1932)
	435.54	35-65 min	death	Nuckolls (1935)
	256	5 min	deep narcosis	Booth & Bixby (1932)
	213.5	2 h	loss of coordination, unconsciousness	Nuckolls (1935)
	213.5	2 h	death within 2 h	Caujolle (1964)
	106.75	2 h	seizures, loss of balance	Nuckolls (1935)
	98.21			
	51.24	2 h	dyspnoea, stupor	Nuckolls (1935)

The 10-min EC_{50} for the CNS effects described was 490 g/m³ for rats (Clark & Tinston, 1982). Signs in rabbits were similar to those in rats, namely incoordination and other signs of CNS depression, followed by respiratory depression and asphyxiation (Sakata et al., 1981). The primary toxic effect of the single inhalation exposure was central nervous system depression, which occurred at very high exposure levels.

Cardiac and pulmonary effects are described in section 7.8.

7.2 Short-term inhalation exposure

In this monograph, short-term exposures are defined as those involving repeated daily exposures for up to 90 days and long-term exposures as those lasting more than 90 days.

Table 7. Acute inhalation toxicity of HCFC 22

Species	Concentration (g/m^3)	Exposure period (min)	Effects observed	Reference
Mouse	1295	120	death (MLC)	Karpov (1963)
	970	30	incoordination, deep narcosis, death (LC_{50})	Sakata et al. (1981)
Rabbit	1050	30	incoordination, cyanosis, death (MLC)	Sakata et al. (1981)
Rat	2100	2	depressed heart rate and blood pressure, ECG changes	Pantaleoni & Luzi (1975a,b)
	1225	15	deep narcosis, death (LC_{50})	Clark & Tinston (1982)
	1050	120	incoordination, accelerated respiration, deep narcosis, death (MLC)	Weigand (1971)
	875	240	death (MLC)	NIOSH (1976)
	775	240	death (LC_{50})	Litchfield & Longstaff (1984)
	700	120	incoordination, accelerated respiration, narcosis	Weigand (1971)
Rat and guinea-pig	1400	120	incoordination, narcosis	Weigand (1971)
Guinea-pig	1050	120	narcosis	Weigand (1971)
	700	2	lacrimation, stupor, tremor	Nuckolls (1940)
Dog	2450	90	partial narcosis, depressed respiration, death	Poznak & Artusio (1960)
	1400	90	incoordination, partial narcosis	Poznak & Artusio (1960)
Monkey	700	5	depressed respiration, heart rate, and blood pressure	Aviado & Smith (1975)

MLC = Minimum lethal concentration.

Cardiac and pulmonary effects are described in section 7.8.

7.2.1 *HCFC 21*

Weigand (1971) exposed five rats, five guinea-pigs, two beagle dogs, and two cats to HCFC 21 (42.7 g/m^3) for

3.5 h/day, 5 days/week, for 4 weeks. The behaviour of all animals remained normal throughout the experiment. The increase in body weight of rats was somewhat retarded and the guinea-pigs lost weight. The blood and urine analyses were normal. Gross pathological examination revealed alterations in the livers of all the guinea-pigs and cats, and one of the dogs, but not of the rats. Histopathological examination showed hepatic single cell necrosis and fatty degeneration in all exposed animals.

Kelly (1976, 1977) and Trochimowicz et al. (1977) exposed a group of 10 male rats to HCFC 21 (42.7 g/m^3) 6 h per day, 5 days/week, for 2 weeks. There were no deaths but the rats lost weight and exhibited marked anaemia and increased serum transaminase levels, indicating liver damage. Pathological examination immediately after the last exposure showed liver necrosis; this necrosis was still present in a recovery group examined 14 days later.

In a further study (Kelly, 1977) groups of 27 male and 27 female Charles River albino rats and 4 male beagle dogs were exposed to HCFC 21 levels of 4.27 or 21.35 g/m^3 6 h per day, 5 days/week, for 90 days. Rats were severely affected; between days 59 and 90, 37% of the rats exposed to the low and 29% exposed to the high concentration died. Standard clinical chemistry investigations showed alterations in liver function. Histopathological examination revealed extensive liver cirrhosis. Neither the mortality nor the histopathological damage was related to the dosage. None of the dogs died; the only significant effects occurred at 21.35 g/m^3 and consisted of slight weight loss during exposure and minimal unspecified morphological changes in the liver.

Lindberg (1979) exposed four groups of Charles River albino rats (35 males and 35 females in each group) to HCFC 21 concentrations of 0, 0.213, 0.64, and 2.13 g/m^3 for 6 h/day, 5 days/week, for 90 days. At a level of 2.13 g/m^3 the average body weight gain was lower than in controls during the early phase of the experiment. Leucocyte counts were elevated in animals exposed to the highest concentration, as were serum alkaline phosphatase and alanine aminotransferase activities. Urine volumes showed a tendency to increase. At this exposure level an increase in urine fluoride concentration was observed in both sexes after 45 days of treatment. There was a similar increase in urinary fluoride concentration at 90 days, with essentially no difference between rats exposed to 0.64 and 2.13 g/m^3. This suggests that saturation of the metabolic transformations occurs at exposure levels at or below 0.64 g/m^3. Histopathological evaluation revealed portal cirrhosis of the liver, interstitial oedema of the pancreas, and degeneration of the seminiferous epithelium at all dose levels. The liver toxicity and indications for metabolism suggest a similar mechanism of toxicity to that demonstrated for trichloromethane (DFG, 1986). Thus, a reactive metabolite may be responsible for the induction of liver effects.

7.2.2 HCFC 22

No effects were seen on body weight, haematological parameters, urine analysis, organ weights or macroscopic and microscopic appearance of the tissues in rats, guinea-pigs, dogs or cats exposed to HCFC 22 (175 g/m^3) 3.5 h per day, 5 days/week, for 4 weeks (Weigand, 1971).

Lee & Suzuki (1981) exposed two groups of 16 male Sprague-Dawley rats to HCFC 22 (0 or 175 g/m^3), 5 h/day for 8 weeks, after which 6 rats in each group were killed and blood and tissue samples taken for haematological and biochemical assays and for histopathological examination. The remaining animals were retained for a fertility study, the results of which are presented in section 7.5. No signs of toxicity were apparent in the exposed animals, and body weight was not affected. Prostate weight was decreased slightly but not the weights of other organs. No histopathological lesion was related to exposure in any of the organs examined. Plasma glucose and triglyceride levels were reduced and plasma cholesterol slightly raised, but no haematological parameter was affected.

Leuschner et al. (1983) exposed Sprague-Dawley rats (35 and 17.5 g/m^3) and beagle dogs (17.5 g/m^3 only) to HCFC 22 6 h/day for 13 weeks. The treatment and control groups consisted of 20 male and 20 female rats and 3 male and 3 female dogs. Investigations of behaviour, body weight, haematology, clinical biochemistry, and organ weights were carried out in both species, and dogs were also subjected to ECG measurements and to an examination of circulatory function. The clinical biochemistry examinations included assays for serum alanine transaminase, aspartate transaminase, and alkaline phosphatase activities, as well as liver function tests. Histopathological examinations were undertaken on a wide variety of tissues. No changes were found in any of these examinations. It was therefore concluded that the no-observed-effect level for HCFC 22 was in excess of 35 g/m^3 in the rat and in excess of 17.5 g/m^3 in the dog.

In a limited experiment on rabbits that also received sodium barbital in the drinking-water, Van Stee & McConnell (1977) found that exposure to HCFC 22 (210 g per m^3, 5 h/day, 5 days/week for 8-12 weeks) induced cardiac arrhythmia in one of 14 rabbits. In addition, some rabbits (number unknown) showed slight histopathological liver damage and a modest elevation in the level of unspecified serum enzymes. The lack of detail and the fact that no controls were used make meaningful conclusions impossible.

7.2.3 Mixed exposure

Two unweaned beagle puppies (one male, one female) weighing 1.5-3 kg inhaled a mixture of HCFC 22 and HCFC 21 (60% : 40%) for 5 min, twice daily, 5 days a week for 2 weeks, at a concentration of 1714 mg/kg body weight. After 1-2 min the puppies became sedated and ataxic, but they recovered a few minutes after removal from exposure. No other effects were noted during investigations that included blood chemistry and urine analyses, and gross and microscopic examinations of the lungs (Knox-Smith & Case, 1973).

7.3 Skin and eye irritation; sensitization

7.3.1 Skin irritation

HCFC 21 produced mild irritation when applied at concentrations higher than 25% in propylene glycol to the shaved, intact skin of guinea-pigs. No irritation was observed at a concentration of 2.5% (Goodman, 1975).

Quevauvillier et al. (1964) reported that a 10-second spray of HCFC 22 on the shaved belly of the rat twice a day, 5 days/week, for 6 weeks caused reddening of the skin and slight swelling of the surface. There was also a delay in hair regrowth. A more recent report (Atochem, 1986) classified the compound as a skin irritant but this effect was only observed following the application of 0.5 ml in liquified form under occlusion to the intact and abraded skin of rabbits.

The irritation induced by a chemical with a boiling point at or near room temperature appears to be related to rapid evaporation, resulting in a drying effect on the skin or mucous membrane.

7.3.2 Eye irritation

Undiluted liquid HCFC 21, chilled to the temperature of dry ice, was placed into the right conjunctival sacs of two rabbits. After 20 seconds, the treated eye of one rabbit was washed with 0.9% saline for 1 min. Slight corneal opacity, transient congestion of the iris and moderate conjunctival irritation in the unwashed eye was seen. There was slight corneal opacity and moderate conjunctival irritation but no iris involvement in the washed eye. Both eyes were normal within 5 days (Brittelli, 1975). This report did not distinguish between the effects of cold, including that caused by evaporation, and the intrinsic properties of HCFC 21. In another study, HCFC 21 was sprayed directly into the eyes of each of six rabbits from a distance of 5 cms. No corneal or iris injury was seen but lacrimation was observed in four of the rabbits examined 1 and 4 h after exposure (Hood, 1964a). Two additional studies on rabbits evaluated the eye irritation potential of diluted HCFC 21 in various solvents (Hood, 1964b; Eddy, 1970). In both studies, 0.1 ml of the test solution was instilled in one eye of each of six albino rabbits. The other eye, which was treated with vehicle, served as a control. The eyes were not washed after treatment. HCFC 21, as either a 50% solution in mineral oil or a 40% solution in propylene glycol or dimethyl phthalate, produced varying degrees of injuries to the cornea, iris, and conjunctivae. Milder irritant effects were seen with a 15% solution. All these effects disappeared within 7 to 10 days.

HCFC 22 was reported to be a slight irritant when the corneas of albino rabbits were exposed to the gas for 5 or 30 seconds (Atochem, 1986).

7.3.3 Skin sensitization

No evidence of skin sensitization with HCFC 21 was found in guinea-pigs by Hood (1964b) or Goodman (1975).

The skin-sensitizing potential of HCFC 22 was tested in 10 male and 10 female Hartley albino guinea-pigs using a modification of the Magnusson-Kligman maximization text. On day 0, Freund Complete Adjuvant was injected intradermally and 0.25 ml of liquified compound was applied topically to the skin under a capsule for 28 h. On days 2, 4, 7, 9, 11, and 14, 0.5 ml of liquified HCFC 22 was applied topically at the same place for 48 h under occlusion. After a period of 2 weeks, the challenge exposure was performed on day 28 at the opposite side of the body: 0.25 ml of liquified HCFC 22 (maximum non-irritating dose as determined in the previous experiment) was applied under a capsule for 48 h. No cutaneous sensitizing reaction was observed during a macroscopic or histological evaluation of the skin 1, 6, 24, and 48 h after removal of the occlusive patch (Atochem, 1986).

7.4 Long-term inhalation exposure

No data are available on the chronic toxicity of HCFC 21 based on exposures longer than 90 days.

Karpov (1963) exposed rats, mice, and rabbits to HCFC 22 (50 g/m^3), as well as rats and mice to 7 g per m^3, for 6 h/day, 6 days/week, over a 10-month period. Body weights, oxygen consumption, CNS function, and biochemical and haematological parameters were recorded, and histopathological examinations of some tissues were undertaken at the end of the test. Depressed body weight gain in mice after 4-6 months, depressed oxygen consumption in rats, CNS function changes in rats and mice, decreased haemoglobin concentration in rabbits, and histopathological (dystrophic) changes in the liver, lungs, and nervous tissue were observed at 50 g/m^3. No effect was seen at the lower exposure level (7 g/m^3).

In a life-time study, Tinston et al. (1981a) exposed Alderley Park Swiss-derived mice (80 male and 80 female per group) to HCFC 22 (0, 3.5, 35, and 175 g/m^3) 5 h per day, 5 days/week, for up to 83 weeks (females) and 94 weeks (males), at which time there was 80% mortality. At week 38, 10 mice per group were killed in order to perform blood assays, including red and white blood cell counts, platelet counts, prothrombin and kaolin-cephalin times, and bone marrow examination. Measurements of plasma ASAT (EC 2.6.1.1) and ALAT (2.6.1.2) activities and urine analyses were also undertaken. The only consistent finding was hyperactivity observed in the male mice exposed to 175 g/m^3. There were no treatment-related effects on mortality or body weight gain. No abnormalities were observed during haematological, biochemical or histopathological investigations, with the exception of the neoplasia described in section 7.7.

Tinston et al. (1981b) performed a similar lifetime study on Alderley Park Wistar-derived rats using the same group sizes and exposure levels. The study lasted 118 weeks in females and 131 weeks in males (80% mortality). Some animals were killed at week 52. The same investigations were carried out as in the case of the mouse study. No clinical abnormalities, increased mortality, or haematological or biochemical changes could be attributed to HCFC 22 at any exposure level. At the highest level (175 g/m^3), there was a decrease in body weight gain in males and increased liver, kidney, and adrenal and pituitary gland weights in the females. A number of non-neoplastic lesions were observed histologically in all the groups but there was no evidence of exposure-relatedness.

7.5 Reproduction, embryotoxicity, and teratogenicity

7.5.1 Reproduction

No data are available on the effects of HCFC 21 on reproduction, except for a limited study by Aranjina (1972) who claimed a decrease in the levels of DNA and total nucleic acids in the liver, brain, ovaries, and placenta of female rats exposed to 0.153 or 0.303 g/m^3 for the whole gestation period. The biological significance of this finding cannot be evaluated due to the lack of adequate reporting.

Lee & Suzuki (1981) tested HCFC 22 for effects on male reproduction in Sprague-Dawley rats. A group of 16 male rats was exposed to a concentration of 175 g/m^3, 5 h per day, for 8 weeks, and a control group of the same size was exposed to filtered air. The animals were examined and weighed weekly. At the end of the 8-week period, six rats from each group were killed, organ weights determined, and histopathological, clinical, chemical, and haematological examinations were carried out. The prostate glands were assayed for fructose and acid phosphatase (EC 3.1.3.2) activity. Immediately after the final exposure, blood was collected from the remaining 10 rats in each group, and the plasma was assayed for follicle stimulating hormone (FSH) and luteinizing hormone (LH). These animals were then used for serial mating, each male being housed with a virgin female for 7 days, after which time the female was replaced with another virgin female; this regime was followed for 10 weeks. Nine days after removal, each female was killed, and the numbers of corpora lutea, total implants, live implants, resorption sites, and dead implants were determined. There was no sign of any overt toxicity. The major organs of treated animals, including testes and epidymes, did not differ in weight from those of the controls. There was a slight decrease in weight of the prostate and coagulating gland in the treated rats but there were no accompanying histological changes. Prostatic fructose and acid phosphatase (EC 3.1.3.2) levels, as well as FSH and LH levels, were not different from those in controls. Serum cholesterol levels were slightly higher, and glucose and triglyceride levels were slightly lower in treated rats than in controls. Overall, there was no significant effect upon the fertility of male rats.

7.5.2 Embryotoxicity and teratogenicity

7.5.2.1 HCFC 21

When Kelly et al. (1978) exposed pregnant rats to 42.7 g/m^3 (6 h/day on days 6-15 of gestation), no clinical signs of toxicity were observed but the rats gained substantially less weight than the control animals. HCFC 21 interfered with the process of implantation: 15 of 25 mated rats had no implants or viable fetuses. The outcome of pregnancy and the fetal development in the other

10 rats were not affected. No teratogenic activity was observed.

7.5.2.2 HCFC 22

Three inhalation teratology studies with HCFC 22 were carried out in 1977 and 1978 (Culik et al., 1976; Culik & Crowe, 1978). Groups of 20-40 pregnant Charles River CD rats were exposed to several concentrations ranging from 0.35 to 70 g/m^3 for 6 h/day on days 4-13 or 6-15 of gestation. There was no evidence of maternal or overt fetal toxicity. The only teratogenic abnormalities observed were small statistically insignificant increases in microphthalmia or anophthalmia. The incidence of microphthalmia/anophthalmia in the three studies is presented in Table 8.

Table 8. Incidence (fetuses/litters) of microphthalmia or anophthalmia in three preliminary studies of HCFC 22 in Charles River CD rats[a]

Study	Exposure levels (g/m^3)						
	0	0.350	1.05	1.75	3.5	35	70
1	0/21				1A/22	2A/21	
2	0/34			1M/33	(2M+/33)		1A/35
3	0/38	1M/40	0/35			2M/34	

[a] From: Culik et al. (1976) and Culik & Crowe (1978)
A = anophthalmia; M = microphthalmia (no quantitative procedures were applied to the assessment of microphthalmia); + = two fetuses were affected in the same litter.

The comparatively low incidence of microphthalmia or anophthalmia in the above study led Palmer et al. (1978a) to conduct a large study designed to improve the sensitivity of the investigation of these infrequent malformations. Female CD rats were exposed to concentrations of 0, 0.35, 3.5 or 175 g/m^3 for 6 h/day on days 6-15 of gestation. Nineteen batches of time-mated females were used, each batch consisting of 34 control rats and 22 rats in each of the three test groups; more than 6000 control fetuses and 4000 fetuses from exposed animals were examined. The animals were killed on day 20 of pregnancy and examined macroscopically. The ovaries were examined for numbers of corpora lutea and the uteri for live young and embryonic fetal death; litter and mean pup weights were recorded. The heads of all fetuses were sectioned and examined with particular reference to microphthalmia, anophthalmia, and associated anomalies. Maternal body weights in the group exposed to 175 g/m^3 were slightly lower than in controls in most batches; the body weight gain was consistently lower in this group than in the controls during the first day of exposure. No other change in the dams was observed. Overall, there was no effect on litter size, post-implantation loss or litter weight. Mean fetal weight was slightly but consistently lower in the group exposed to 175 g/m^3 than in controls, but not at the two lower levels of exposure. The number of fetuses and the incidence of litters with anophthalmia or microphthalmia are shown in Table 9. There was no significant difference from controls with respect to the incidence of anophthalmia/microphthalmia at low or intermediate exposure levels. At the highest exposure level, there was a small but significant ($P < 0.05$) increase in the incidence of litters containing fetuses with total eye malformations and anophthalmia. No other gross fetal abnormality was found.

Palmer et al. (1978b) also carried out a teratology study in rabbits. New Zealand white rabbits were exposed to 0, 0.35, 3.5, or 175 g/m^3 for 6 h/day on days 6-18 of pregnancy. There were 14-16 pregnant females per group. The animals were killed on day 29 of pregnancy for an assessment of litter data and an examination of fetuses for major malformations, minor anomalies, and variants. There were no significant treatment-related effects in females, and pregnancies were normal. Maternal body weight gain was slightly but consistently lower in the animals exposed to 175 g/m^3 during the first 4 days of exposure but, thereafter, weight gain was comparable to that in the controls. Litter size, post-implantation loss, and litter and mean fetal weights were unaffected. There was no significant increase in the incidence of major or minor fetal abnormalities.

Table 9. The number and incidence of fetuses with eye malformations (anophthalmia and microphthalmia) in rats exposed to HCFC 22[a]

HCFC 22 concentration (g/m³)	Eye malformations		Anophthalmia		Microphthalmia	
	No. of fetuses affected	Incidence per 1000 litters	No. of fetuses affected	Incidence per 1000 litters	No. of fetuses affected	Incidence per 1000 litters
0	3	4.94	1	1.65	2	3.29
0.35	5	12.66	1	2.53	4	10.13
3.5	3	7.69	1	2.56	2	5.13
175	10	26.11[b]	6	15.67[a]	4	10.44

[a] From: Palmer et al. (1978a).
[b] Statistically significant ($P < 0.05$) by a one-sided stratified contingency chi-square test

7.6 Mutagenicity

7.6.1 HCFC 21

HCFC 21 was not mutagenic when incubated for 72 h with *Salmonella typhimurium* (TA98, TA100, TA1535, TA1537, TA1538) with or without metabolic activation. The compound was not mutagenic to *Saccharomyces cerevisiae* D4 (Brusick, 1976).

7.6.2 HCFC 22

The data from *in vitro* and *in vivo* studies are summarized in Table 10. HCFC 22 induced mutations in *Salmonella typhimurium* base-pair substitution strains TA1535 and TA100 but not strains TA98, TA1537 or TA1538 after gaseous exposure.

The response was independent of the presence or absence of a metabolizing system, as would be expected from the lack of metabolism in animals. Negative responses were reported in mutation assays using *Schizosaccharomyces pombe* and *Saccharomyces cerevisiae* (Loprieno & Abbondandolo, 1980), plant cells (Van't Hoft & Schairer, 1982), Chinese hamster cells (McCooey, 1980; Loprieno & Abbondandolo, 1980), and in a cell transformation assay (BHK21; Loprieno & Abbondandolo, 1980).

Anderson et al. (1977a) found an increase in chromosomal damage in rats exposed to 3.5 g/m^3, 6 h/day, for 5 days. However, there was less evidence of chromosomal damage at exposures of 35 and 525 g/m^3. Similar results were observed following a single 2-h exposure using the same concentrations. Anderson & Richardson (1979) repeated the experiment using lower exposure levels (0.035, 0.35, 1.75, and 3.5 g/m^3). There was an increase in chromosomal damage, but again it was not dose related. Both of these studies were reviewed by Litchfield & Longstaff (1984) and Longstaff (1988). Loprieno & Abbondandolo (1980) did not find chromosomal changes in the bone marrow of mice in a study in which HCFC 22 was administered by gavage.

There was no evidence for dominant lethality in a study on male Sprague-Dawley rats exposed to 177 g/m^3,

Table 10. The genetic toxicity of HCFC 22

Assay	Organism	Species/strain/cell type	Metabolic activation[a]	Testing conditions	Results	Reference
Reverse mutation	bacteria	*Salmonella typhimurium* TA1535; TA1537; TA1538; TA98; TA100	±	20 & 40% gas for 6 h	negative	Barsky (1976)
Reverse mutation	bacteria	*Salmonella typhimurium* TA1535; TA1537; TA98; TA100	±	up to 40% gas for 48 h	positive for TA1535 only, activation independent	Koops (1977)
Reverse mutation	bacteria	*Salmonella typhimurium* TA1535; TA1538; TA98; TA100	±	incubated with 50% gas for 24 h	positive for TA1535, (TA100 positive in 1 of 3 experiments), activation independent	Longstaff & McGregor (1978)
Reverse mutation	bacteria	*Salmonella typhimurium* TA100	±	50% gas for 24 h	positive	Bartsch et al. (1980)
Reverse mutation	bacteria	*Salmonella typhimurium* TA1535; TA1538; TA98; TA100	±	50% for 24 h	positive for TA1535 and TA100, activation independent	Longstaff et al. (1984)
Forward mutation	yeast	*Schizosaccharomyces pombe*	±	20 mM solution generated at 500 ml/min of 50% gas	negative	Loprieno & Abbondandolo (1980)
Cell mutation	plant	*Tradescantia*	N/A	closed chamber, highest ineffective dose 1.16 g/m^3	negative	Van't Hoft & Schairer (1982)
Cell mutation	hamster	Chinese hamster ovary cells	±	tested at 20-92% gas	negative	McCooey (1980)

Table 10 (contd).

Assay	Organism	Species/strain/cell type	Metabolic activation[a]	Testing conditions	Results	Reference
Cell mutation	hamster	Chinese hamster lung V 79 cells	±	20 mM solution generated at 500 ml/min of 50% gas	negative	Loprieno & Abbondandolo (1980)
Cell mutation (host mediated)	yeast	Schizosaccharomyces pombe or Saccharomyces cerevisiae in CD-1 mice	N/A	816 mg/kg body weight in corn oil by gavage	negative	Loprieno & Abbondandolo (1980)
Gene conversion	yeast	Saccharomyces cerevisiae	±	20 mM solution generated at 500 ml/min of 50% gas	negative	Loprieno & Abbondandolo (1980)
Unscheduled DNA synthesis	human	heteroploid EUE cell line	±	20 mM solution generated at 500 ml/min of 50% gas	negative	Loprieno & Abbondandolo (1980)
Cell transformation	hamster	BHK-21 cells	-	20 mM solution generated at 500 ml/min of 50% gas	negative	Loprieno & Abbondandolo (1980)
Chromosome aberrations in vivo	mouse	CD-1, bone marrow cells	N/A	816 mg/kg body weight in corn oil by gavage	negative	Loprieno & Abbondandolo (1980)
Chromosome aberrations in vivo	rats 8/group	bone marrow cells	N/A	inhalation for 2 h at 3.5, 35, or 525 g/m^3	dose/related increase of chromosome damage, significant only at high dose	Anderson et al. (1977a)

Table 10 (contd).

Chromosome aberrations in vivo	rats 8/group	bone marrow cells	N/A	inhalation at 3.5, 35, or 525 g/m^3, 6 h/day for 5 days	significant increase of chromosome aberrations at low and mid-dose but not at high dose	Anderson et al. (1977)
Chromosome aberrations in vivo	mouse	CD-1, bone marrow cells	N/A	816 mg/kg body weight in corn oil by gavage	negative	Loprieno & Abbondandolo (1980)
Dominant lethal	mouse 20/group	CD-1	N/A	inhalation for 6 h/day, 5 days, at 3.5, 35, or 350 g/m^3	positive in some parameters, not time or dose related	Hodge et al. (1979)
Dominant lethal	mouse 20/group	CD-1	N/A	inhalation for 6 h/day, 5 days, at 0.035, 0.35, 1.75, 3.5, 35, 175 g/m^3	positive in some parameters, not time or dose dependent	Hodge et al. (1979)
Dominant lethal	rat	Sprague-Dawley	N/A	175 g/m^3, 5 h/day, for 8 weeks	negative	Lee & Suzuki (1981)
Micronucleus in vivo	mouse	bone marrow cells	N/A	175 and 525 g/m^3 for 6 h	negative	Howard et al. (1989)

^{a} ± indicates that separate experiments were carried out with and without metabolic activation; NA = not applicable

5 h/day, for 8 weeks (Lee & Suzuki, 1981). In two dominant lethal assays on mice at exposure levels of 0.035-350 g/m³, there were statistically significant differences from control values in certain parameters (e.g., early fetal death, reduction of fertility) at various points (Anderson et al., 1977b; Hodge et al., 1979). However, there was no time or exposure-relatedness. In addition, results were not reproducible.

In an inhalation study in mice, Howard et al. (1989) found no evidence of micronucleus induction at exposure levels at 175 and 525 g/m³.

Most of these studies have been reviewed by Litchfield & Longstaff (1984) and Longstaff (1988). With the exception of positive findings in mutation assays using specific strains of *Salmonella* (TA1535 and TA100), HCFC 22 did not show activity in microorganisms or in mammalian *in vitro* and *in vivo* systems. These included mutation, unscheduled DNA synthesis assays *in vitro*, and cytogenetic and dominant lethal assays in two species of rodents. Overall, the available information does not indicate a genotoxic effect of HCFC 22 in mammalian systems.

7.7 Carcinogenicity

No data are available on the carcinogenicity of HCFC 21.

In a life-time study, Tinston et al. (1981b) exposed groups of 80 male and 80 female Alderley Park Wistar-derived rats to HCFC 22 (0, 3.5, 35, or 175 g/m³, 5 h per day, 5 days/week) for 118 weeks in females and 131 weeks in males (the period by which mortality had reached approximately 80%). No treatment-related clinical abnormalities, increased mortality, or haematological or biochemical changes were observed. The only effects were a body weight reduction in males exposed to 175 g/m³ and increased weight of liver, kidney, and adrenal and pituitary glands in females. In males there was no increase in the number of benign tumours, but there was a slight increase in the number of rats with malignant tumours at the highest exposure level (Table 11). This increase was primarily due to the increased incidence of rats with fibrosarcoma. The only organ that was consistently associated

Table 11. Tumour incidence in male rats exposed to HCFC 22 in a lifetime inhalation study[a]

	HCFC concentrations (g/m^3)			
	0[b]	3.5	35	175
Total malignant tumours	16/80 18/80	27/80	22/80	33/80
Selected sites				
Fibrosarcomas	5/80 7/80	8/80	5/80	18/80[c]
Salivary gland fibrosarcomas	1/80 0/80	1/80	0/80	7/80[c]
Zymbal gland tumours	0/80 0/80	0/80	0/80	4/80[c]

[a] From: Tinston et al. (1981b).
[b] Two control groups were used.
[c] Statistically significant difference in relation to the control group ($P < 0.05$)

with this increase was the salivary gland, but, according to Litchfield & Longstaff (1984), this may have been the consequence of generalized subcutaneous fibrosarcomas developing in a submandibular site and involving the salivary gland only by chance. The increase in fibrosarcomas was observed only in the late stages of the study (between 105 and 130 weeks). No significant increase was found at lower exposure levels. Four male rats in the group exposed to 175 g/m^3 were found to have Zymbal gland tumours at the highest exposure level. Female rats did not exhibit such changes at any of the exposure levels.

In a further study, Tinston et al. (1981a) exposed groups of 80 male and 80 female Alderley Park Swiss-derived mice to HCFC 22 (0, 3.5, 35, or 175 g/m^3, 5 h per day, 5 days/week) for up to 83 weeks in males and 94 weeks in females (approximately 80% mortality). The only finding was hyperactivity. There was no significant increase in the incidences of benign or malignant tumours in treated male or female mice compared to the controls, with the exception of a small increase in the incidence of hepatocellular carcinomas in the males at the highest exposure level. However, the incidences of both benign and malignant nodules were within the range of historical control values for this strain of mouse.

There have been two well-documented and well-conducted lifetime inhalation studies by Tinston et al. (1981a,b) in rats and mice. These did not indicate any increased tumour incidence except for fibrosarcomas in male rats at the highest exposure level of 175 g/m^3. The lack of tumour response was also seen in another inhalation study using a different mouse strain and a maximum exposure level of 17.5 g/m^3, and in a limited gavage study in rats.

The overall results of these studies indicate no tumorigenic effect up to exposure levels of 35 g/m^3. It has been suggested that the increased tumorigenic response in male rats at the high exposure level may be due to the presence of HCFC 31, a known mutagen and animal carcinogen, as an impurity (DFG, 1986), but more evidence is needed to confirm this suggestion.

Maltoni et al. (1982, 1988) exposed groups of 60 male and 60 female Sprague-Dawley rats and Swiss mice by inhalation to HCFC 22 (concentrations of 0, 3.5, and 17.5 g/m^3, for 6 h/day, 5 days/week) for 104 weeks (rats) or 78 weeks (mice). There were no differences in body weights or survival between treated and control animals (details not given). No increases in tumour incidence were observed.

Longstaff et al. (1984) administered to a group of 36 male and a group of 36 female Alderley Park Wistar-derived rats HCFC 22 (300 mg/kg body weight) in corn oil by gavage daily (on 5 days/week) for 52 weeks. One similarly sized control group received no treatment and two control groups were dosed with corn oil. The study was terminated at week 125. The treatment had no effect on body weight gain or mortality. No increased incidence of tumours was observed in any of the organs in the treated groups compared either with untreated controls or those given corn oil.

7.8 Special studies - cardiovascular and respiratory effects

Chlorofluorocarbons have long been known to sensitize the heart to adrenaline-induced arrhythmias (Reinhardt et al., 1971; Zakhari & Aviado, 1982). Respiratory effects have also been noted.

7.8.1 HCFC 21

A summary of data on the effects of HCFC 21 on cardiovascular and respiratory function is given in Table 12.

Table 12. Cardiovascular and respiratory function studies on HCFC 21 in different animal species

Species	Concentration (g/m^3)	Effect	Reference
Mouse	427	arrhythmia (with and without exogenous adrenaline)	Aviado & Belej (1974)
Rat	811.3	apnoea, cardiac arrest	Friedman et al. (1973)
Cat	1700	sensitization to exogenous adrenaline, cardiac arrhythmia	Branch et al. (1990)
Dog	42.7	lowest concentration causing cardiac sensitization (with exogenous adrenaline)	Mullin (1975)
	106.75	EC$_{50}$ for cardiac sensitization (with exogenous adrenaline)	Clark & Tinston (1973)
	106.75	tachycardia, bronchoconstriction	Aviado (1975a,b)
	106.75	rise in pulmonary resistance, decrease in pulmonary compliance	Belej & Aviado (1975)
	427	hypotension	
Monkey	106.75	minimal concentration for induction of cardiac arrhythmia, tachycardia, myocardial depression, and hypotension	Aviado (1975); Belej et al. (1974)
	106.75	reduction in respiratory minute volume	Aviado & Smith (1975)
	213.5	bronchodilation with depression of respiration	Aviado (1973, 1975)

In non-anaesthetized dogs, the concentration of HCFC 21 leading to cardiac arrhythmias in 50% of catecholamine-treated dogs (EC$_{50}$ value) was calculated to be 107 g/m^3 for a 5-min exposure period (Clark & Tinston, 1973). In a separate study, the lowest dose causing this effect was 43 g/m^3 (Mullin, 1975). Tachycardia, bronchoconstriction, and loss of pulmonary compliance were also found at exposure levels of 107 g/m^3 when anaesthetized dogs were given a 7-min exposure (Aviado, 1975a,b; Belej & Aviado, 1975).

In anaesthetized monkeys (species not further defined), spontaneous cardiac arrhythmias, tachycardia, and hypertension, as well as reduced respiratory minute volume, were found following exposure to 107 g/m^3 for 7 min (Belej et al., 1974; Aviado & Smith, 1975). In contrast to dogs, bronchodilation and early respiratory depression were found in monkeys exposed to 214 g/m^3 (Aviado, 1975a,b).

Spontaneous arrhythmias occurred in mice exposed to 425 g/m^3 for 7 min (Aviado & Belej, 1974). Brachycardia, apnoea, and cardiac arrest, accompanied by lowered respiratory rates and raised lung tidal volumes, were apparent in rats exposed to 811 g/m^3 (Friedman et al., 1973).

These effects of HCFC 21 are also characteristic of the fully halogenated chlorofluorocarbons (see WHO, 1990).

7.8.2 HCFC 22

A summary of data on the cardiovascular and respiratory effects of HCFC 22 in experimental animals is presented in Table 13.

Belej et al. (1974) evaluated the cardiotoxic potential of HCFC 22 in three Rhesus monkeys anaesthetized with sodium pentobarbitone. The trachea of each animal was canulated and the chest was open for direct measurement of cardiac function. Following exposure to 350 or 700 g/m^3 for 5 min, various indices of cardiac function were measured, i.e. heart rate, myocardial force, aortic blood pressure, and left atrial pressure. The only changes found were a slight but significant depression of myocardial contractility and a drop in aortic blood pressure in monkeys exposed to either concentration.

The effect of HCFC 22 on respiratory function was studied by Aviado & Smith (1975) in three Rhesus monkeys anaesthetized by intravenous application of sodium phenobarbital (30 mg/kg) and exposed for 5 min to 350 g/m^3 through a canulated trachea. Electrocardiograms and fem-

Table 13. Cardiovascular and respiratory function studies of HCFC 22 in different animal species

Species	Concentration (g/m^3)	Duration of exposure	Effects	Reference
Rat	525-2100	2 min	decreased heart rate and changes in ECG	Pantaleoni & Luzi (1975a)
	1050-2100	2 min	decreased myocardiac contractility, ECG changes, arterial hypotension	Pantaleoni & Luzi (1975b)
Mouse	700	6 min	no arrhythmias with or without exogenous adrenaline;	Aviado & Belej (1974)
	1400	6 min	arrhythmias seen only with exogenous adrenaline	
Rabbit	210	5 h/day for 8-12 weeks	one of 14 rabbits developed arrhythmias; also received phenobarbital (no controls were used)	Van Stee & McConnell (1977)
Dog	1750	5 min	lowest concentration causing cardiac sensitization (with exogenous adrenaline)	Mullin (1975)
	87.5	5 min	no effects (with exogenous adrenaline)	Reinhart et al. (1971)
	175	5 min	cardiac sensitization (with exogenous adrenaline)	
	490	5 min	EC_{50} for cardiac sensitization (with exogenous adrenaline)	Clark & Tinston (1982)
	17.5	6 h/day for 90 days	no effect on ECG and circulatory functions	Leuschner et al. (1983)
Monkey	350	5 min	depression of myocardial contractility;	Belej et al. (1974)
	700	5 min	decreased aortic blood pressure	
	350	5 min	no effect on pulmonary measurement	Aviado & Smith (1975)
	700	5 min	slightly increased pulmonary resistance	

oral arterial blood pressures were recorded, and pulmonary airway resistance and compliance were estimated from measurements of tracheal airflow and transpulmonary pressure. No significant change was recorded at 350 g/m^3, but a slight, yet significant, elevation in pulmonary resistance was observed at 700 g/m^3.

Arrhythmias were also noted in cats given 1700 g per m^3 of 40% HCFC 22 for 10 minutes (Branch et al., 1990).

Reinhardt et al. (1971) assessed the ability of HCFC 22 to induce cardiac sensitization to adrenaline in male beagle dogs exposed to 87.5 or 175 g/m^3 via a face mask, using 12 dogs at each exposure level. After 5 min of exposure, a challenge injection of adrenaline (0.008 mg/kg) was given. At 87.5 g/m^3 no effect was noted, but at 175 g/m^3 two animals displayed cardiac sensitization as demonstrated by serious arrhythmias. The authors classified HCFC 22 as a weak cardiac sensitizing agent. This is consistent with the finding of Mullin (1975) that the minimum concentration of HCFC 22 at which cardiac sensitization occurs in dogs injected with adrenaline is 175 g/m^3.

Leuschner et al. (1983) did not observe any effect on ECG and circulatory functions in dogs exposed to 17.5 g per m^3 (6 h/day, 7 days/week) for 90 days.

Clark & Tinston (1982) determined the exposure levels of HCFC 22 causing acute cardiac sensitization after adrenaline treatment in dogs and mice. The concentration affecting 50% of the animals (EC_{50}) was 490 g/m^3 for both species.

Aviado & Belej (1974) exposed anaesthetized Swiss mice to 700 or 1400 g/m^3 for 6 min via a face mask. In the first experiment no adrenaline was given, whereas in the second 6 µg/kg was administered intravenously. Only at the higher exposure level with adrenaline were arrhythmias recorded.

Pantaleoni & Luzi (1975a,b) found a decreasing heart rate and ECG changes in rats exposed to 525-2100 g/m^3 for 2 min. In a second study, rats exposed to 1050-2100 g/m^3 for 2 min showed a decrease in cardiac contractile strength, followed by a decrease in carotid pressure, ECG changes, and arterial hypotension. Vagotomy partially inhibited the appearance of the ECG changes. In both experiments the modified parameters returned to normal within 2 min of breathing normal air.

When Van Stee & McConnell (1977) exposed 7 male and 7 female rabbits to 210 g/m^3 for 5 h/day (5 days/week, for 8-12 weeks), 1 female rabbit, which was also receiving phenobarbital, developed arrhythmia, probably of supraventricular origin. No controls were used in this limited study.

8. EFFECTS ON HUMANS

8.1 General population exposure

8.1.1 Accidents

One case of acute poisoning has been reported. A young boy was found dead in a small room with his mouth around the nozzle of a tank containing HCFC 22. No details were provided (Garriot & Petty, 1980).

8.1.2 Controlled human studies

Ten healthy volunteers and ten patients suffering from a pronounced arterial hypoxaemia due to broncho-pulmonary disease were exposed by inhalation to an aerosol mixture containing 60% HCFC 21 and 40% CFC 11. They inhaled 202 ml of the aerosol within 2.5 h or 126 ml during 10 successive breaths. No ECG changes were observed in either group (Fabel et al., 1972).

A group of eleven subjects, of whom seven were maintenance technicians of large cooling and refrigerating systems, were exposed to chlorofluorocarbons at weighted exposure levels similar to their usual occupational exposures. They were also exposed experimentally for 130 min to HCFC 22 concentrations of 0.71 or 1.89 g/m^3, to a mixture of HCFC 115 (4 g/m^3) and HCFC 22 (1.4 g/m^3), or to HCFC 115 (23.4 g/m^3) and HCFC 22 (10.5 g/m^3). No significant changes in ventilatory lung function or cardiac function were observed (Valic et al., 1982).

8.2 Occupational exposure

A fisherman working in the hold of a fishing boat, and another man who attempted to rescue him, died when the hold was filled with HCFC 22 from a broken refrigerant pipe. A judicial autopsy, performed 5 days after the accident, revealed that the blood of both cadavers was dark red. The lungs showed remarkable congestion, oedema, haemorrhage, emphysema and pigment-laden macrophages in the alveoli, which were detected histologically. Fine fatty droplets were observed in the cytoplasm of liver cells. No other signs that might have caused injury or sudden death were noted. HCFC 22 was detected by gas chromatography in samples of various tissues, blood, urine, and pericardial fluid. From these results, the cause of death was concluded to be oxygen deficiency resulting from the filling of the ship's hold with HCFC 22 from the broken refrigerant pipe (Haba & Yamamoto, 1985).

Speizer et al. (1975) carried out a questionnaire survey on the incidence of palpitations in a group of hospital pathology laboratory workers (118 people, 94% of total staff in the pathology department) who used HCFC 22 in the preparation of frozen sections. This study was triggered by the death of one of the workers from myocardial infarction. A 3.5-fold excess of palpitations was reported amongst exposed workers compared to an unexposed control group from the radiology department (85 people, 93% of total staff in the department). Using the number of slides produced as an indicator of HCFC 22 exposure, the authors reported a dose-response relationship. The only exposure data provided were breathing-zone samples collected from two subjects during a 2-min period of HCFC 22 use; these revealed concentrations of 1062 mg per m^3. The possible role of exposure to other substances was not taken into consideration, the control group was not adequate, and the validity of making estimates of exposure on the basis of the number of frozen sections prepared is questionable.

Antti-Poika et al. (1990) did not find a clear connection between cardiac arrhythmia and exposure to a mixture of HCFC 22 and CFC 12 (average concentrations were 170-815 ppm HCFC 22 and 202 ppm CFC 12). One subject, however, had several ventricular ectopic beats; his average level of exposure was 170 ppm but the peak concentration was as high as 3200 ppm.

A case of peripheral neuropathy in a commercial refrigeration repair worker prompted a survey of a group of 27 refrigeration workers (Gunter et al., 1982; Campbell et al., 1986). No case of peripheral neuropathy was identified, and chest radiograms, pulmonary function tests, electrocardiograms, and blood and urine analysis results were all within normal limits. A questionnaire completed by all subjects indicated that light-headedness and palpitations were more common in the refrigeration workers than in the control group of 14 unexposed workers. Personal air samples collected over a "typical" shift showed levels of HCFC 22 of 5.1 mg/m^3, and co-exposure to CFC 115 and CFC 12 was assumed.

In a survey by Edling & Ohlson (1988), 89 workers with intermittent exposure to chlorofluorocarbons were examined during their work with refrigerant equipment. The refrigerants used were mainly CFC 12 (in 56% of the cases) and HCFC 22 (in 32% of cases), the rest being CFC 11, CFC 500, and CFC 502 (a mixture of CFC 115 and HCFC 22). There was no statistically significant difference in ECG results between periods with and without exposure nor was there any exposure-related trend when workers were grouped in exposure groups. No effect on the central nervous system was found. Edling et al. (1990) confirmed that the data did not indicate that these fluorocarbons induced cardiac arrhythmia among occupationally exposed refrigerator repair workers.

In a preliminary report, death rates were investigated in 539 workers occupationally exposed to chlorofluorocarbons CFC 12, HCFC 22, and HCFC 502. No increase in total deaths (18 cases) was found among those employed for more than 6 months when compared with the expected number (26) derived from national statistics. There were five deaths due to heart disorders (compared to 9.6 expected), six deaths due to cancer (versus 5.7 expected), and two deaths from lung cancer (versus 1.0 expected). No significant increase was found even in those exposed for more than 3 or 10 years (Szmidt et al., 1981). In view of the limited data, no conclusions could be drawn from this study.

9. EFFECTS ON OTHER ORGANISMS IN THE LABORATORY AND FIELD

No information is available on the effects of HCFCs 21 or 22 on organisms in the environment.

10. EVALUATION OF HUMAN HEALTH RISKS AND EFFECTS ON THE ENVIRONMENT

10.1 Evaluation of human health risks

10.1.1 Direct health effects resulting from exposure to partially halogenated chlorofluorocarbons

No information on exposure levels for the general population has been reported. Tropospheric concentration measurements range from 4.3 to 86 ng/m^3 (1-20 ppt) for HCFC 21 and from 110 to 326 ng/m^3 (31-92 ppt) for HCFC 22.

Information on simulated aerosol propellant use indicates peak HCFC 22 exposures ranging from 5 to 8 g/m^3. These levels rapidly (in 10-20 seconds) decline to background values of 0.025 to 0.045 g/m^3. No information is available regarding exposure concentrations at the workplace, but in a simulated beauty-salon model an 8-h time-weighted average HCFC 22 exposure level of 0.09-0.125 g per m^3 was measured.

The recommended occupational exposure limits in different countries range from 1800 to 3540 mg/m^3 (500-1000 ppm) for HCFC 22 and from 40 to 45 mg/m^3 (approximately 10 ppm) for HCFC 21.

On the basis of available information, the kinetics and metabolism of HCFC 21 and HCFC 22 are characterized by rapid pulmonary absorption and distribution throughout the body. After cessation of exposure these chemicals are cleared from the blood within minutes. Metabolic transformation of HCFC 22 is minimal, if it occurs at all. Therefore, toxic effects of metabolites are extremely unlikely. However, there is evidence for some metabolism of HCFC 21, based on increased urinary excretion of fluoride after exposure of animals.

The acute toxicity of both chemicals is very low. As with the fully halogenated chlorofluorocarbons, it is characterized in animals by effects on the central nervous and cardiopulmonary systems at inhalation concentrations of 175 g/m^3 or more.

In animal studies with repeated inhalation exposure, HCFC 21 is more toxic than HCFC 22. The latter causes little or no toxicity at exposure levels likely to be of any practical significance to humans. In various animal species HCFC 21 induces liver damage, which might be due to the formation of one or more reactive metabolites.

The carcinogenicity of HCFC 21 has not been studied. However, the substance is not genotoxic in bacterial and yeast systems. In one lifetime study, exposure to HCFC 22 led to small, but statistically significant, increases in the incidence of fibrosarcomas in the salivary gland region and Zymbal's gland, but only in male rats at high exposure levels (175 g/m^3). Other lifetime studies in rat and mouse did not produce any indication of carcinogenicity, and HCFC 22 was not genotoxic in mammalian systems. Thus, HCFC 22 is unlikely to constitute a carcinogenic risk to humans at environmental or controlled occupational exposure levels.

An embryotoxicity study with HCFC 21 in rats revealed implantation losses at maternally toxic doses (42.7 g per m^3), but there were no teratogenic effects. No further information is available on the potential effects of HCFC 21 on reproduction. At the maternally toxic high HCFC 22 exposure concentration of 175 g/m^3, a small increase in the incidence of eye malformations was reported in rats. In view of these findings and the absence of any effects in rabbits at similar exposure levels, HCFC 22 is considered unlikely to pose a risk to developmental processes in humans at environmental or controlled occupational exposure levels. No effects were observed in a one-generation reproduction study on rats exposed to HCFC 22 at a concentration of 175 g/m^3 (5 h/day for 8 weeks).

In summary, the toxic potency of HCFC 21 is greater than that of HCFC 22. This is reflected in the 50- to 100-fold lower occupational exposure limits for HCFC 21 than for HCFC 22. Controlled occupational exposures are unlikely to represent a significant risk to humans. Although high peak exposure levels may occur during consumer use of products containing HCFC 22 as a propellant, the exposure period is short and so adverse health effects are unlikely. Environmental exposure levels are extremely low and are not considered likely to cause direct effects to human health.

10.1.2 Health effects expected from a depletion of stratospheric ozone by partially halogenated chlorofluorocarbons

The possible indirect health effects (e.g., an increase in the incidence of skin cancer and immunotoxic and ocular effects) of fully halogenated chlorofluorocarbons, resulting from an increase in UV-B radiation due to a depletion of the ozone layer, have been discussed in Environmental Health Criteria 113. The ozone-depleting potential (ODP) of HCFC 22 is about 20 times lower than for CFC 11 and that of HCFC 21 is estimated to be lower still. Based on this information, the indirect health effects of these chemicals are expected to be considerably lower than those of the fully halogenated chlorofluorocarbons.

10.2 Effects on the environment

No information is available to evaluate the direct ecological effects posed by HCFC 21 and HCFC 22. With respect to the indirect "greenhouse" effect, HCFC 22 is calculated to have a third to a quarter of the "global-warming potential" of the fully halogenated CFC 11, while that of HCFC 21 is lower still.

11. CONCLUSIONS AND RECOMMENDATIONS FOR PROTECTION OF HUMAN HEALTH AND THE ENVIRONMENT

11.1 Conclusions

The available toxicological data on HCFC 21 and HCFC 22 reviewed in this monograph show differences in their toxicological potential. HCFC 22 exhibits a low acute and chronic toxicity and the data indicate no mutagenic and only a low carcinogenic potential in animals. Developmental effects only occur at high and maternally toxic exposure concentrations. Although HCFC 21 also has a low acute toxicity potential, it appears to be more toxic than HCFC 22, causing damage mainly to the liver following repeated exposure.

Owing to the observed differences in the toxicity profiles, different occupational exposure limits have been recommended for these two chemicals. Based on available exposure information, it is concluded that these chemicals are unlikely to elicit adverse health effects in humans associated with consumer and environmental exposures and workplace exposures at or below current occupational limits. The human health risks are mainly confined to those following accidental exposures and might also result from occupational exposure if this is not properly controlled.

HCFC 21 and HCFC 22 have a lower ozone-depleting potential than the fully halogenated chlorofluorocarbons and should therefore pose a lower indirect health risk.

The global-warming potentials of HCFC 21 and HCFC 22 are considerably lower than those of the fully halogenated chlorofluorocarbons.

11.2 Recommendations for protection of human health and the environment

1. Since the toxicity of HCFC 22 is low and the ozone-depleting and global-warming potentials are lower than those of the fully halogenated chlorofluorocarbons, it can be considered as a transient substitute for the

chlorofluorocarbons included in the Montreal Protocol. However, in line with the conclusions of the 1990 London Meeting of the Montreal Protocol Parties, efforts should be maintained to develop substitutes that would pose negligible or no risk to the environment and to develop alternative technologies.

2. Although HCFC 21 poses a low risk to the environment, it is not recommended as a substitute for the chlorofluorocarbons included in the Montreal Protocol because of its liver toxicity.

12. FURTHER RESEARCH

1. In order to fill in gaps in the available information on general population exposure and global environmental effects, it is recommended that:

    -   the concentrations of HCFCs in food packaging materials and in packaged food should be measured;
    -   monitoring of atmospheric levels of HCFCs in different parts of the world should be maintained;
    -   more studies of heterogeneous atmospheric phenomena over polar regions should be conducted in order to substantiate the values for the ozone-depleting potential of HCFCs.

2. HCFC 21 is used on a very limited scale and it is not likely to become a substitute for the chlorofluorocarbons included in the Montreal Protocol. If its use is expected to increase substantially in the future, more information on potential health effects are needed, such as:

    -   chronic toxicity in animals, with special reference to hepatotoxicity;
    -   the mechanism of action in relation to liver toxicity;
    -   genotoxic and carcinogenic potential;
    -   results of targeted health surveillance in populations under exposure.

3. In view of the interim use of HCFC 22, the existing gaps in knowledge concerning its effects on human health and the environment should be filled. The following areas of research are recommended:

    -   a study of the mechanism of action in order to clarify the genotoxic and carcinogenic effects;
    -   targeted health surveillance in populations subject to elevated levels of exposure to this substance.

13. PREVIOUS EVALUATIONS BY INTERNATIONAL BODIES

An evaluation of the carcinogenicity of HCFC 22 by the International Agency for Research on Cancer (IARC, 1987) was reported as follows:

"There is inadequate evidence of carcinogenicity in humans; there is limited evidence of carcinogenicity in experimental animals. The agent cannot be classified as to its carcinogenicity to humans (group 3)."

REFERENCES

ACGIH (1990) Threshold limit values for chemical substances and physical agents and biological exposure indices 1990-1991. Cincinnati, Ohio, American Conference of Governmental Industrial Hygienists.

American Chemical Society (1985) Chemcyclopedia 85. Washington, DC, American Chemical Society.

Antonova VI, Suhov JZ, Salmina ZA, & Petrova NA (1983) [Toxicology of Freon 22 and its permissible water level.] Gig i Sanit **8**: 69-70 (in Russian).

Anderson D & Richardson CR (1979) Arcton 22: A second cytogenetic study in the rat. Alderley Park, Cheshire, Imperial Chemical Industries Ltd, Central Toxicology Laboratory (Report No. CTL/P/445).

Anderson D, Richardson CR, Howard C, Riley RA, & Wright TM (1977a) Arcton 22: Cytogenetic study in the rat. Alderley Park, Cheshire, Imperial Chemical Industries Ltd, Central Toxicology Laboratory (Report No. CTL/R/429).

Anderson D, Hodge MCE, Weight TM, & Riley C (1977b) Arcton 22: Dominant lethal study in the mouse. Alderley Park, Cheshire, Imperial Chemical Industries Ltd, Central Toxicology Laboratory (Report No. CTL/R/430).

Antti-Poika M, Heikilla J, & Saarinen L (1990) Cardiac arrhythmias during occupational exposure to fluorinated hydrocarbons. Br J Ind Med **47**: 138140.

Aranjina T (1972) [Effects of aliphatic hydrocarbons and fluorinated and chlorinated derivatives on the content of nucleic acids in animal tissues during embryogenesis.] Permsk Gos Med Inst **110**: 69-71 (in Russian).

Atkinson R (1985) Kinetics and mechanisms of the gas-phase reactions of the hydroxyl radical with organic compounds under atmospheric conditions. Chem Rev **85**(1): 169-201.

ATOCHEM (1986) Chlorofluoromethane: local tolerance tests in the rabbit, ocular irritation - cutaneous irritation - cutaneous sensitizing test in the guinea-pig. Hazleton IFT (Report No. 603344).

Aviado DM (1975a) Toxicity of propellants. In: Proceedings of the 4th Annual Conference on Environmental Toxicology. Ohio, Wright-Patterson AFB, Aerospace Medical Research Laboratory, pp 291-329.

Aviado DM (1975b) Toxicity of aerosol propellants in the respiratory and circulatory systems. X. Proposed classification. Toxicology **3**: 321-332.

Aviado DM & Belej MA (1974) Toxicity of aerosol propellants on the respiratory and circulatory systems. I. Cardiac arrhythmia in the mouse. Toxicology **2**: 31-42.

Aviado DM & Smith DG (1975) Toxicity of aerosol propellants on the respiratory and circulatory systems. Respiration and circulation in primates. Toxicology **3**(2): 241-252.

Bales RE (1978) Fluorocarbons - workers exposure in four facilities. Washington, DC, National Technical Information Service, p 22 (PB 297772).

Barsky FC (1976) In vitro microbial mutagenicity studies of methane, chlorodifluoro. Wilmington, Delaware, Du Pont de Nemours and Co., Haskell Laboratory (Report No. 398-76 E).

Bartsch H, Malaveille C, Camus AM, Martel-Planche G, Brun G, Hautefeuille A, Sabadie N, Barbin A, Kuroki T, Drevon C, Piccoli C, & Montesano R (1980) Validation and comparative studies on 180 chemicals with *S. typhimurium* strains and V79 Chinese hamster cells in the presence of various metabolizing systems. Mutat Res 76(1): 1-50.

Belej MA & Aviado DM (1975) Cardiopulmonary toxicity of propellants for aerosols. J Clin Pharmacol 15: 105-115.

Belej, MA, Smith DG, & Aviado DM (1974) Toxicity of aerosol propellants in the respiratory and circulatory systems. IV. Cardiotoxicity in the monkey. Toxicology 2(4): 381-395.

Booth MS & Bixby EM (1932) Fluorine derivatives of chloroform. Ind Eng Chem 24: 637.

Bouraly M & Lemoine JC (1988) An assessment of the exposure to Forane 22 (HCFC 22) used as a propellant in aerosols. Paris, ATOCHEM (Report No. 12-550).

Bower FA (1973) Nomenclature and chemistry of fluorocarbon compounds. Springfield, Virginia, US National Technical Information Service, p 9 (NTIS AD Report No. 751423).

Branch CA, Ewing JR, Fagan SC, Goldberg DA, & Welch KMA (1990) Acute toxicity of a nuclear magnetic resonance cerebral blood flow indicator in cats. Stroke 21(8): 1172-1177.

Britelli MR (1975) Eye irritation test in rabbits. Wilmington, Delaware, Du Pont de Nemours and Co., Haskell Laboratory (Report No. 751-75).

Brunner F, Crescentini G, Mangani F, Brancaleoni E, Capielo A, & Ciccioli P (1981) Determination of halocarbons in air by gas chromatography-high resolution mass spectrometry. Anal Chem 53: 798-801.

Brusick DJ (1976) Mutagenicity evaluation of Genetron R 21. Kensington, Maryland, Litton Bionetics (LBI Project No. 2683).

Campbell DD, Lockey JE, Petajan J, Gunter BJ, & Rom WN (1986) Health effects among refrigeration repair workers exposed to fluorocarbons. Br J Ind Med 43(2): 107-111.

Carney I (1977) Arcton 22 (chlorodifluoromethane) relationship between blood levels and inhaled concentrations in anaesthetised artificially ventilated rats. Alderley Park, Cheshire, Imperial Chemical Industries Ltd, Central Toxicology Laboratory (Report No. CTL/R/422).

References

Caujolle F (1964) Comparative toxicity of refrigerants. Inst Int Froid 1: 21-54.

CFR (1981) Petition to remove chlorodifluoromethane from the list of toxic pollutants under Section 307 (a) of the Clean Water Act. Fed Reg 46: 2276.

Clark DG & Tinston DJ (1973) Correlation of the cardiac sensitizing potential of halogenated hydrocarbons with their physiochemical properties. Br J Pharmacol 49: 355-357.

Clark DG & Tinston DJ (1982) Acute inhalation toxicity of some halogenated and non-halogenated hydrocarbons. Hum Toxicol 1(3): 239-247.

Crescentini G & Bruner F (1979) Evidence for the presence of Freon 21 in the atmosphere. Nature (Lond) 279: 311-312.

Culik R & Crowe CD (1978) Embryotoxic and teratogenic studies in rats with inhaled chlorofluoromethane (FC-22). Third study. Wilmington, Delaware, Du Pont de Nemours and Co., Haskell Laboratory (Report No. 314-78).

Culik R, Kelly DP, & Burgess BA (1976) Embryotoxic and teratogenic studies in rats with inhaled chlorofluoromethane (F-22). Wilmington, Delaware, Du Pont de Nemours and Co., Haskell Laboratory (Report No. 970-76).

DFG (Deutsche Forschungsgemeinschaft) (1986) [Materials hazardous to health. Justification of maximum workplace concentrations from the toxicological and occupational health viewpoint (12th instalment).] Weinheim, VCH-Verlagsgesellschaft mbH (in German).

DFG (Deutsche Forschungsgemeinschaft) (1990) [Maximum workplace concentrations and tolerance values for biological materials.] Weinheim, VCH-Verlagsgesellschaft mbH (in German).

Eddy CW (1970) Eye irritation test in rabbits. Wilmington, Delaware, Du Pont de Nemours and Co., Haskell Laboratory (Report No. 423-70).

Edling C & Ohlson CG (1988) [Health risks with exposure to freons.] Uppsala, Sweden, Department of Occupational Medicine, University Hospital (in Swedish).

Edling C, Ohlson CG, Ljungkvist G, Oliv A, & Soederholm B (1990) Cardiac arrhythmia in refrigerator repairmen exposed to fluorocarbons. Br J Ind Med 47: 207-212.

Fabel MR, Wettengel R, & Hartman W (1972) [Myocardial ischemia and arrhythmias in the application of pressurized aerosols in man.] Dtsch Med Wochenschr 97: 428-431 (in German).

Fischer DA, Hales ChH, Wang WCh, Ko MKW, & Sze ND (1990a) Model calculations of the relative effects of CFCs and their replacements on global warming. Nature (Lond) 344: 513-916.

Fischer DA, Ko M, Wuebbles D, & Isaksen I (1990b) Reply to Solomon & Tuck (1990). Nature (Lond) 348: 203-204.

FRG (1990) [CFC-halon prohibition ordinance.] Bonn, Lower House of Parliament, pp 11-19 (Report 11/8166) (in German).

Friedman SA, Cammarato M, & Aviado DM (1973) Toxicity of aerosol propellants on the respiratory and circulatory systems. II. Respiratory and bronchopulmonary effects in the rat. Toxicology 1: 345-355.

Garber W-D (1989) The greenhouse effect. In: Responsibility means doing without - How to rescue the ozone layer. Berlin, Umweltbundesamt, pp 30-33.

Garriot J & Petty CS (1980) Death from inhalant abuse: toxicological and pathological evaluation of 34 cases. Clin Toxicol 16: 305-315.

Goodman NC (1975) Primary skin irritation and sensitization test on guinea-pigs. Wilmington, Delaware, Du Pont de Nemours and Co., Haskell Laboratory (Report No. 750-75).

Grasselli JG & Ritchey WM, ed. (1975) CRC atlas of spectral data and physical constants for organic compounds. Cleveland, Ohio, CRC Press, vol 3, p 592.

Grayson M (1978) Kirk-Othmer encyclopedia of chemical technology. New York, Wiley & Sons.

Gunter BJ, Campbell DD, Rom WW, & Petajan JH (1982) Health hazard evaluation of refrigeration workers, Salt Lake City, Utah. Cincinnati, Ohio, National Institute for Occupational Safety and Health (HETA-81-043-1207).

Haba K & Yamamoto H (1985) Two cases of death caused by Freon 22 gas in fish hold. Res Pract Forensic Med. 28: 103-108.

Hammit JK, Camm F, Conell PS, Mooz WE, Wolf KA, Wuebbles DJ, & Bamezai A (1987) Future emission scenarios for chemicals that may deplete stratospheric ozone. Nature (Lond) 330: 711-716.

Hanhoff-Stemping I (1989) Properties which affect the environment. In: Responsibility means doing without - How to rescue the ozone layer. Berlin, Umweltbundesamt, pp 33-52.

Hansch C & Leo A (1979) Substituent constants for correlation analysis in chemistry and biology. New York, John Wiley & Sons, p 172.

Hartop PJ & Adams MG (1989) Simulated consumer exposure to propellant HCFC 22 (chlorofluoromethane) on aerosol personal products. Int J Cosmet Sci 11: 27-34.

Hawley GG (1981) The condensed chemical dictionary, 10th ed. New York, Van Nostrand Reinhold, p 236.

Hodge MCE, Anderson D, Bennett IP, & Weight TM (1979) Arcton 22: Second dominant lethal study in the mouse. Alderley Park, Cheshire, Imperial Chemical Industries Ltd, Central Toxicology Laboratory (Report No. CTL/R/450 - revised).

Hood DB (1964a) Eye irritation tests. Wilmington, Delaware, Du Pont de Nemours and Co., Haskell Laboratory (Report No. 107-64).

References

Hood DB (1964b) Eye irritation test. Wilmington, Delaware, Du Pont de Nemours and Co., Haskell Laboratory (Report No. 105-64).

Horrath AL (1982) Halogenated hydrocarbons. New York, Marcel Dekker, p 48.

Howard CA, Cryer N, & Greenwood M (1989) Arcton 22: An evaluation in the mouse micronucleus test. Alderley Park, Cheshire, Imperial Chemical Industries Ltd, Central Toxicology Laboratory (Report No. CTL/R/2433).

Höfler F, Schneider J, & Moeckel HJ (1986) Analytical investigations on the composition of gaseous effluents from a garbage dump. I. General data, sampling technique, and analysis of low binding halogenated compounds. Fresenius Z Anal Chem 325: 365-368.

IARC (1987) Overall evaluations of carcinogenicity: An updating of IARC monographs 1-42. Lyon, International Agency for Research on Cancer (IARC Monographs on the Evaluation of Carcinogenic Risk to Humans, Suppl 7).

Karpov BD (1963) [The chronic toxicity of Freon-22.] Tr Leningrad Sanit Gig Med Inst 75: 231-240 (in Russian).

Kelly DP (1976) Two-week inhalation toxicity studies. Wilmington, Delaware, Du Pont de Nemours and Co., Haskell Laboratory (Report No. 149-76).

Kelly DP (1977) Ninety-day inhalation exposure of rats and dogs to vapours of dichlorofluoromethane (F-21). Wilmington, Delaware, Du Pont de Nemours and Co., Haskell Laboratory (Report No. 493-77).

Kelly DP, Culic R, Trochimowicz HJ, & Fayerweather WF (1978) Inhalation teratology studies on three fluorocarbons. Toxicol Appl Pharmacol 45: 293 (Abstract 170).

Khalil MAK & Rasmussen RA (1981) Increase of $CHCl_2$ in the earth's atmosphere. Nature (Lond) 292: 823-824.

Khalil MAK & Rasmussen RA (1983) Gaseous tracers of arctic haze. Environ Sci Technol 17: 157-164.

Knox-Smith J & Case MT (1973) Subacute and chronic toxicity studies of fluorocarbon propellants in mice, rats and dogs. Toxicol Appl Pharmacol 26(3): 438-443.

Komoriya H, Nakamura I, & Ohya I (1980) A toxicological study of the effects of Freon 22 inhalation - the behaviour of rats exposed to Freon inhalation and an evaluation of Freon concentrations in their tissue. Jpn J Leg Med 42(4-5): 372-380.

Koops A (1977) Mutagenic activity of methane, chlorodifluoro in the *Salmonella* microsome assay. Wilmington, Delaware, Du Pont de Nemours and Co., Haskell Laboratory (Report No. 577-77).

Krüger BC & Fabian P (1986) Model calculations of the reduction of atmospheric ozone by different halogenated hydrocarbons. Ber Bundenges Phys Chem 90: 1062-1066.

Lee IP & Suzuki K (1981) Studies in the male reproductive toxicity of Freon 22. Fundam Appl Toxicol 1(3): 266-270.

Leifer R, Sommers K, & Guggenheim SF (1981) Atmospheric trace gas measurements with a new clean air sampling system. Geophys Res Lett **8**: 1079-1082.

Leuschner F, Neumann BM, & Huebschner F (1983) Report on subacute toxicological studies with several fluorocarbons in rats and dogs by inhalation. Arzneimittelforschung **33**: 1475-1476.

Lindberg DC (1979) Subacute inhalation toxicity study with Genetron 21 in Albino rats. Decatur, Illinois, Industrial Biotest Laboratories.

Litchfield MH & Longstaff E (1984) The toxicological evaluation of chlorofluorocarbon 22 (CFC 22). Food Chem Toxicol **22**(6): 465-475.

Long GR & Bialkowski SE (1985) Saturation effects in gas-phase photochemical deflection spectrophotometry. Anal Chem **57**: 1079-1083.

Longstaff E (1988) Carcinogenic and mutagenic potential of several fluorocarbons. Ann NY Acad Sci **534**: 283-298.

Longstaff E & McGregor D (1978) Mutagenicity of a halocarbon refrigerant monochlorfluoromethane (E-22) in *Salmonella typhimurium*. Toxicol Lett **2**(1): 1-4.

Longstaff E, Robinson M, Bradbrook C, Styles JA, & Purchase IFH (1984) Genotoxicity and carcinogenicity of fluorocarbons. Assessment of short-term *in vitro* tests and chronic exposure in rats. Toxicol Appl Pharmacol **72**(1): 15-31.

Loprieno N & Abbondandolo A (1980) Comparative mutagenic evaluation of some industrial compounds. In: Northop KH & Garner RC ed. Short-term tests systems for detecting carcinogens. Berlin, Springer Verlag, p 333.

McCooey KT (1980) Methane, dichlorodifluoro-, Chinese hamster ovary cell assay for mutagenicity. Wilmington, Delaware, Du Pont de Nemours Co., Haskell Laboratory (Report No. 149-90).

Makide Y & Rowland FS (1981) Tropospheric concentrations of methylchloroform, CH_3CCl_3, in January 1978 and estimates of the atmospheric residence times for hydrohalocarbons. Proc Natl Acad Sci (USA) **78**: 5933-5937.

Maltoni C, Ciliberti A, & Carretti D (1982) Experimental contributions in identifying brain potential carcinogens in the petrochemical industry. Ann NY Acad Sci **381**: 216-249.

Maltoni C, Lefemine G, Tovoli D, & Perino G (1988) Long-term carcinogenicity bioassay on three chlorofluorocarbons (trichlorofluoromethane, FC11; dichlorodifluoromethane, FC12; chlorodifluoromethane, FC22) administered by inhalation to Sprague-Dawley rats and Swiss mice. Ann NY Acad Sci **534**: 261-282.

Molina MJ, Rowland FS, Chou CC, Smith WS, Ruiz NV, Crescentini G, & Milstein R (1976) Atmospheric chemistry of several chlorofluorocarbon compounds. Fluorocarbons 21, 22, 31, 13, 113, 114, 115. Abstracts from the 12th International Symposium on Free Radicals, Laguna Beach, California, USA.

Morita M, Miki A, Kazama H, & Sakata M (1977) Case report of deaths caused by Freon gas. Forensic Sci 10(3): 253-260.

Mullin LS (1975) Cardiac sensitization. Wilmington, Delaware, Du Pont de Nemours and Co., Haskell Laboratory (Report No. 707-75).

NASA (1988) Present state of knowledge of the upper atmosphere 1988, an assessment report, based on a report of the Ozone Trend Panel. Washington, DC, National Aeronautics and Space Administration (NASA Reference Publication No. 1208).

NIOSH (1976) In: Christensen HE & Fairchild FJ ed. Registry of toxic effects of chemical substances. Cincinnati, Ohio, National Institute for Occupational Safety and Health, p 690 (Entry No. PA 73900).

NIOSH (1985) Dichlorofluoromethane. In: NIOSH manual of analytical methods. Cincinnati, Ohio, National Institute for Occupational Safety and Health (Method 2516).

Nuckolls AH (1935) The comparative life, fire and explosion hazard, common refrigerants. Chicago, Illinois, The Underwriter's Laboratory (Miscellaneous Hazards No. 2630).

Nuckolls AH (1940) The toxicology of some commercial fluorocarbons. In: Proceedings of the 2nd Annual Conference on Environmental Toxicology, Fairborn, Ohio. Chicago, Illinois, The Underwriter's Laboratories (Miscellaneous Hazards No. 3134).

Palmer AK, Cozens DD, Clark R, & Clark GC (1978a) Effect of Arcton 22 on pregnant rats: relationship to anophthalmia and microphthalmia. Huntingdon, United Kingdom, Huntingdon Research Centre (Report No. 174/7820).

Palmer AK, Cozens DD, Clark R, & Clark GC (1978b) Effect of Arcton 22 on pregnancy of the New Zealand white rabbit. Huntingdon, United Kingdom, Huntingdon Research Centre (Report No. 177/78505).

Pantaleoni GC & Luzi V (1975a) [Cardiotoxicity of monochlorodifluoromethane (contribution to the study on the toxicopathology of sniffing syndrome). Note 3.] Rass Med Sper 22(6): 265-269 (in Italian).

Pantaleoni GC & Luzi V (1975b) [Cardiotoxicity of monochlorodifluoromethane (contribution to the study on the toxicopathology of sniffing syndrome). Note 4.] Rass Med Sper 22(6): 270-274 (in Italian).

Pellizzari ED, Hartwell TD, Harris B, Waddell RD, Whitaker DA, & Erickson MD (1982) Purgeable organic compounds in mother's milk. Bull Environ Contam Toxicol 28(3): 322-328.

Penkett SA, Prosser NJD, Rasmussen RA, & Khalil MAK (1980) Measurements of $CHFl_2$ in background tropospheric air. Nature (Lond) 286: 793-795.

Peter H, Filser JG, Szentpaly LV, & Wiegand HJ (1986) Different pharmacokinetics of dichlorodifluoromethane (CFC 21) and chlorodifluoromethane (CFC 22). Arch Toxicol 58(4): 282-283.

Poznak AV & Artusio JF (1960) Anesthetic properties of a series of fluorinated compounds. I. Fluorinated hydrocarbons. Toxicol Appl Pharmacol 2: 363-373.

Quevauvillier A, Schrenzel M, & Huyen VN (1964) Tolérance locale (peau, muqueuses, plaies, brûlures) chez l'animal, aux hydrocarbures chlorofluorés. Thérapie 19: 247-263.

Ramanathan V, Cicerone RJ, Singh HP, & Kiehl JT (1985) Trace gas trends and their potential role in climate change. J Geophys Res 90(3): 5547-5566.

Rasmussen RA & Khalil MAK (1983) Natural and anthropogenic trace gases in the lower troposphere of the arctic. Chemosphere 12: 371-375.

Rasmussen RA, Khalil MAK, Penkett SA, & Prosser NSD (1980) Chlorodifluoromethane (F-22) in the earth's atmosphere. Geophys Res Lett 7: 809-812.

Rasmussen RA, Khalil MAK, Crawford AJ, & Fraser OY (1982) Natural and anthropogenic trace gases in the southern hemisphere. Geophys Res Lett 9: 704-707.

Rasmussen RA, Khalil MAK, Crescentini G, Mangani F, Mastrogiocomo AR, & Bruner F (1983) Interlaboratory comparison, preparation and stability of dichlorofluoromethane samples and standards. Anal Chem 55: 1834-1836.

Reinhardt CF, Azar A, Maxfield ME, Smith PE, & Mullin LS (1971) Cardiac arrhythmias and aerosol "sniffing". Arch Environ Health 22: 265.

Sakata M, Kazama H, Miki A, Yoshida A, Haga M, & Morita M (1981) Acute toxicity of fluorocarbon-22: toxic symptoms, lethal concentration and its fate in rabbit and mouse. Toxicol Appl Pharmacol 59(1): 64-70.

Salmon AG, Basu SK, Fitzpatrick M, & Nash JA (1979) Arcton 22: metabolism study *in vitro* in rats and *in vivo*. Alderley Park, Cheshire, Imperial Chemical Industries Ltd, Central Toxicology Laboratory (Report No. CTL/P/438).

Salzburger M, Jacobi HN, & Pahlke G (1989) CFS as refrigerants. In: Responsibility means doing without - How to rescue the ozone layer. Berlin, Umweltbundesamt, pp 104-121.

Sax NI (1984) Dangerous properties of industrial materials, 6th ed. New York, Van Nostrand Reinhold, p 701.

Seigneur C, Caram H, & Carr BW Jr (1977) Atmospheric diffusion and chemical reaction of the chlorofluoromethanes $CHFCl_2$ and CHF_2. Atmos Environ 11: 205-215.

Shimohara K, Sueta S, Tabata T, & Shigemori N (1979) [Determination of monochlorodifluoromethane in the ambient air by electron-capture gas chromatography with a back-flushing system.] Taiki Osen Gakkaishi 14: 31-37 (in Japanese).

Sittig M (1985) Handbook of toxic and hazardous chemicals and carcinogens, 2nd ed. Park Ridge, New Jersey, Noyes Data Corporation, pp 230, 325-326.

References

Smart BE (1980) Fluorine compounds, organic. In: Grayson M & Eckroth D ed. Kirk-Othmer encyclopedia of chemical technology. New York, John Wiley & Sons, vol 10, pp 856-870.

Solomon S & Tuck A (1990) Evaluating ozone depletion potentials. Nature (Lond) 348: 203.

Speizer FE, Wegman DH, & Ramirez A (1975) Palpitation rate associated with fluorocarbon exposure in a hospital setting. New Engl J Med 292: 624.

SRI (1985) Chemical economics handbook. Menlo Park, California, SRI International.

Stoibe RE, Legget DC, Jenkins TE, Mussman RP, & Rose WI (1971) Organic compounds in volcanic gas from Santiaguito volcano, Guatemala. Geol Soc Am Bull 82: 2299-2302.

Szmidt M, Axelson D, & Edling C (1981) [Cohort study of subjects exposed to freon.] Acta Soc Med Suec 90: 77 (in Swedish).

Tappan CH & Waritz RS (1964) Acute inhalation toxicity. Wilmington, Delaware, Du Pont de Nemours and Co., Haskell Laboratory (Report No. 128-064).

Tinston DJ, Chart IS, Godley MJ, Gore CW, Gaskell BA, & Litchfield MH (1981a) Chlorodifluoromethane (CFC 22): long-term inhalation study in the mouse. Alderley Park, Cheshire, Imperial Chemical Industries Ltd, Central Toxicology Laboratory (Report No. CTL/P/547).

Tinston DJ, Chart IS, Godley MJ, Gore CW, Litchfield MH, & Robinson M (1981b) Chlorodifluoromethane (CFC 22): long-term inhalation study in the rat. Alderley Park, Cheshire, Imperial Chemical Industries Ltd, Central Toxicology Laboratory (Report No. CTL/P/548).

Trochimowicz HJ, Lyon JP, Kelly DP, & Chiu T (1977) Ninety-day inhalation toxicity studies on two fluorocarbons. Toxicol Appl Pharmacol 41: 200.

UNEP (1988) Meeting on the Current Status of Atmospheric Science Concerning the Ozone Layer - Ozone Depleting Potential (ODP) Calculations. Nairobi, United Nations Environment Programme (Documents UNEP/OzL.Sc.1/14/ Add. 2, UNEP/OzL.Sc.1/14/Rev. 1 and UNEP/OzL.Wg.Data.2/Inf.1).

UNEP (1990) Second meeting of the Parties to the Montreal Protocol on Substances that Deplete the Ozone Layer. Nairobi, United Nations Environment Programme (Document UNEP/OzL.Pro.2/3).

UNEP/WMO (1989) Scientific assessment of stratospheric ozone: 1989. 4. Halogenated ozone depletion and global warming potentials. Reports prepared for the meeting of the Open-Ended Working Group on the Parties to the Montreal Protocol, Nairobi, United Nations Environment Programme. Nairobi, 28 August-5 September 1989.

US EPA (1981) Petition to remove chlorodifluoromethane from the list of toxic pollutants under Section 307(a) of the Clean Water Act. Fed Reg 46: 2276-2278.

Valic F, Skuric Z, & Zuskin E (1982) [Environmental exposure to freons 12, 22 and 502.] Rad JAZU 402(18): 229-243 (in Croatian with English summary).

Van Stee EW & McConnell EE (1977) Studies of the effects of chronic inhalation exposure of rabbits to chlorodifluoromethane. Environ Health Perspect. 20: 246 (Abstract).

Van't Hoft J & Schairer LA (1982) Tradescantia assay system for gaseous mutagens. A report of the US Environmental Protection Agency Genetox Program. Mutat Res 99: 303-315.

Vidal-Madjar C, Parey F, Excoffier JL, & Bekassy S (1981) Quantitative analysis of chlorofluorocarbons. Absolute calibration of the electron-capture detector. J Chromatogr 203: 247-261.

Weast RC ed. (1985) CRC handbook of chemistry and physics, 66th ed. Boca Raton, Florida, CRC Press, pp C-349, D-211.

Weigand W (1971) Investigations into the inhalation toxicity of fluorine derivatives of methane, ethane and cyclobutane. Zentralbl Arbeitsmed 21(5): 149-156.

WGD (1987) Health based recommended occupational exposure limits for fluorocarbons. The Hague, Directorate-General of Labour, Dutch Expert Committee for Occupational Standards (RA 15/87).

WHO (1990) Environmental Health Criteria 113: Fully halogenated chlorofluorocarbons. Geneva, World Health Organization, 164 pp.

WMO (1986) Global Ozone Research and Monitoring Project. Report No. 16: Atmospheric ozone 1985 - Assessment of our understanding of the processes controlling its present distribution and change. Geneva, World Meteorological Organization.

Woollen BH (1988) Arcton 22: Pharmacokinetics in the pregnant female rat following inhalation exposure. Alderley Park, Cheshire, Imperial Chemical Industries Ltd, Central Toxicology Laboratory (Report No. CTL/R/997).

Zakhari S & Aviado DM (1982) Cardiovascular toxicology of aerosol propellants, refrigerants, and related solvents. In: Van Stee EW ed. Cardiovascular toxicology. New York, Raven Press, pp 281-324.

Zurer P (1989) CFC substitutes. Candidates pass early toxicity tests. Chem Eng News Oct 9: 4.

Zurer P (1990) Fate of CFC alternatives remains up in the air. Chem Eng News July 16: 5-6.

RESUME

1. Identité, propriétés physiques et méthodes d'analyse

Les deux chlorofluorocarbures étudiés dans la présente monographie (le dichlorofluorométhane, HCFC 21, et le chlorodifluorométhane, HCFC 22) sont des hydrofluorocarbures (HCFC), c'est-à-dire des composés obtenus par substitution partielle des atomes d'hydrogène du méthane par des atomes de fluor et de chlore. Seul le HCFC 22 a une importance commerciale. Le HCFC 21 et le HCFC 22 sont des gaz ininflammables (dans des conditions normales de température et de pression). Ils sont incolores et pratiquement inodores. Le HCFC 21 est légèrement et le HCFC 22 moyennement soluble dans l'eau. Tous deux sont miscibles aux solvants organiques. Le HCFC existe sous forme de gaz liquéfié.

Il existe plusieurs méthodes pour doser ces deux composés, notamment la chromatographie en phase gazeuse avec détection par capture d'électrons ou ionisation de flamme, la chromatographie en phase gazeuse couplée à la spectrométrie de masse ou encore la spectrométrie par déflection photothermique.

2. Sources d'exposition humaine et environnementale

Ces deux HCFC n'existent pas dans la nature. Le HCFC 21 n'est produit qu'en petites quantités sans application professionnelle. On estime qu'en 1987, la production mondiale de HCFC 22 était de 246 000 tonnes.

Le HCFC 22 est essentiellement dissipé dans l'environnement lors de la réparation, de l'utilisation et de la mise au rebut de réfrigérateurs ou de climatiseurs. On estime qu'à l'heure actuelle, il se perd chaque année dans le monde environ 120 000 tonnes de HCFC 22. On a également fait état d'émissions accidentelles par des bateaux de pêche.

Le HCFC est utilisé comme réfrigérant, comme intermédiaire dans la production de tétrafluoréthylène et comme agent de soufflage dans la fabrication du polystyrène. On l'utilise également en petites quantités comme gaz propulseur dans les bombes aérosols.

3. Transport, répartition et transformation dans l'environnement

Le logarithme du coefficient de partage octanol/eau du HCFC 22 est de 1,08, ce qui rend la bioaccumulation improbable. On estime que la durée de séjour du HCFC 21 dans la troposphère est d'environ 2 ans et celle du HCFC 22 d'environ 17 ans. Dans la troposphère, la principale voie de dégradation consiste sans doute dans la réaction avec les radicaux OH. Seule une petite fraction de ces deux composés pénètre dans la stratosphère, où, par suite de leur réaction avec des radicaux oxygénés, ils libèrent du chlore qui attaque la couche d'ozone. Toutefois, on pense que le HCFC 22 produit moins de 1% du chlore qui s'attaque à la couche d'ozone. On estime que le potentiel de destruction de l'ozone du HCFC 22 est de 0,05, celui du HCFC 21 étant sans doute plus faible. Quant au potentiel de réchauffement de la planète, calculé par rapport à celui du CFC 11 pris comme unité, on estime qu'il est de 1/3 à 1/4, celui du HCFC 21 étant encore plus faible.

4. Concentrations dans l'environnement et exposition humaine

On ne dispose d'aucune donnée sur les concentrations dans l'eau ni sur la présence de ces composés dans des denrées alimentaires, encore que le HCFC 22 soit utilisé pour la fabrication de récipients en polystyrène expansé destinés au conditionnement de produits alimentaires. Il n'existe pas non plus de données sur l'exposition humaine au HCFC 21, mais deux études relatives à l'utilisation d'aérosols contenant 17-65% de HCFC 22 ont montré que lors d'expositions de courte durée (10-20 s), des concentrations maximales de l'ordre de 5000 à 8000 mg/m^3 pouvaient être atteintes. Les employés d'instituts de beauté pourraient être exposés à des concentrations de 90-125 mg/m^3 (moyenne pondérée par rapport au temps sur 8 h). Cependant, il s'agit là de valeurs qui sont bien inférieures aux limites fixées par la réglementation en vigueur en Allemagne, aux Etats-Unis et aux Pays-Bas.

Le HCFC 22 se dissipe rapidement dans l'atmosphère. On a fait état de concentrations de l'ordre de 326 mg/m^3 en 1986 et il semble que cette valeur augmente de 11% par an.

5. Cinétique et métabolisme chez l'animal de laboratoire et chez l'homme

Il existe quelques données concernant l'absorption, la distribution, le métabolisme et l'excrétion du HCFC 21. Ce composé est très probablement absorbé après inhalation, à en juger d'après les effets généraux et les taux élevés de fluorures dans l'urine que l'on a observés lors d'études toxicologiques sur le rat. Après injection intrapéritonéale à des rats, le HCFC 21 est exhalé et les données cinétiques, de même que l'excrétion des fluorures indiquent que le composé est effectivement métabolisé. Toutefois, on ignore dans quelle proportion le HCFC 21 est transformé et aucun autre métabolite que le fluorure n'a été identifié.

Après inhalation, le HCFC 22 est rapidement et bien absorbé chez le rat, le lapin, et l'homme et il se répartit largement dans l'organisme. De fortes concentrations de HCFC 22 ont été retrouvées dans le sang, le cerveau, le coeur, le foie, les reins et la graisse des viscères chez des lapins morts des suites d'une exposition à ce composé ainsi que dans des échantillons nécropsiques prélevés sur le cerveau, les poumons, le foie et les reins de victimes d'une intoxication par le HCFC 22. L'élimination est rapide, la majeure partie du composé étant excrétée avec une demi-vie de 1 min chez le lapin et de 3 min chez le rat. Chez l'homme, le produit est éliminé en quantité limitée selon un processus triphasique (demi-vie de 3 min, 12 min et 2,7 h respectivement).

Le HCFC 22 réapparaît presque entièrement et sans modification dans l'air exhalé, après avoir été inhalé ou injecté par voie intrapéritonéale à des rats ou à des sujets humains. Il y a de bonnes raisons de penser qu'il ne subit pas de transformation métabolique notable chez le rat *in vivo* ni en présence de préparations de foie de rat.

6. Effets sur les mammifères de laboratoire et les systèmes *in vitro*

On ne possède pas de données satisfaisantes sur les effets toxiques aigus par voie orale de ces deux composés.

Les principaux effets d'une seule inhalation de HCFC 21 ou de HCFC 22 sont en gros les mêmes chez nombre d'espèces animales. Par cette voie, leur toxicité est faible. Les manifestations toxiques observées sont caractéristiques des chlorofluorocarbures, c'est-à-dire qu'elles consistent en une perte de coordination et une narcose. A forte concentration, on voit quelquefois des arythmies cardiaques et des effets pulmonaires (concentrations supérieures ou égales à 106,7 g/m^3).

On a rapporté des cas d'irritation cutanée ou oculaire dus à ces deux composés. On peut toutefois penser que ces effets étaient dus davantage à des pertes thermiques par évaporation qu'aux propriétés chimiques des produits en cause. D'ailleurs, aucun d'eux n'a provoqué de sensibilisation cutanée.

Les seules études consacrées à ces deux composés ne portaient que sur une seule voie d'administration, la voie respiratoire. Les principaux effets observés chez le rat, le cobaye, le chien et le chat consistaient en lésions au niveau du foie. On n'en a pas tiré de valeur pour la dose sans effet observable. Des lésions histopathologiques du foie ont été notées à des doses ne dépassant pas 0,213 g/m^3 administrées 6 h par jour, 5 jours par semaine pendant 90 jours. A ces doses on a également observé un oedème interstitiel du pancréas et une dégénérescence des cellules épithéliales au niveau des tubes séminifères. Ces lésions n'apparaissaient pas lors des études où le HCFC 22 était administré à des doses comprises entre 17,5 g/m^3 (13 semaines) et 175 g/m^3 (4 ou 8 semaines).

Aucune étude de longue durée n'a été consacrée au HCFC 21 chez l'animal. Le seul effet non cancérogène qui ait été systématiquement relevé dans le cas du HCFC 22, consistait dans l'hyperactivité manifestée par des souris mâles exposées à des doses de 175 g/m^3 de ce produit 5 h par jour, 5 jours par semaine pendant toute leur existence.

Il n'y a pas eu d'étude toxicologique classique consacrée aux effets du HCFC 21 sur la fécondité. Lors d'une étude d'embryotoxicité effectuée sur des rats, on n'a pas observé d'effets tératogènes, mais dans un grand nombre de cas, la nidation n'a pas pu s'effectuer. Le HCFC 22 (175 g/m^3, 5 h par jour, 5 jours par semaine pendant 8 semaines) n'a eu aucun effet sur la capacité de reproduc-

Résumé

tion des rats mâles. Un léger surnombre de malformations oculaires sans signification véritable ayant été observé lors de trois études tératologiques chez le rat, on a entrepris une étude de grande envergure afin de voir si le HCFC 22 avait effectivement tendance à provoquer des lésions oculaires. On a ainsi constaté une augmentation faible, mais néanmoins statistiquement significative, du nombre de portées dans lesquelles des foetus présentaient une microphtalmie ou une anophtalmie lorsque les mères étaient exposées à une dose de 175 g/m^3, 6 h par jour, du 6ème au 15ème jour de la gestation. Cette dose était légèrement toxique pour les mères (réduction du poids corporel par rapport aux témoins).

Le HCFC 21 s'est révélé non mutagène dans trois tests de mutagénicité, l'un sur levure et les deux autres sur bactéries. Aucune autre donnée sur ce point n'a pu être obtenue. Le HCFC 22 a présenté une activité mutagène sur modèle bactérien *(S. typhimurium)* mais cette activité ne s'est pas confirmée sur d'autres systèmes bactériens ni sur cellules mammaliennes, que ce soit *in vivo* ou *in vitro*. Ces épreuves portaient sur les mutations géniques, la synthèse non programmée de l'ADN, la cytogénétique des cellules de moelle osseuse et les mutations léthales dominantes (rat et souris).

Seul le HCFC 22 a fait l'objet d'épreuves de cancérogénicité *in vivo*. Deux groupes de chercheurs ont fait inhaler du HCFC 22 à des rats et à des souris pendant toute la durée de leur existence. On n'a observé une surfréquence des tumeurs cancéreuses que dans une seule étude au cours de laquelle des rats mâles avaient été soumis à 175 g/m^3, 5 jours par semaine, pendant des périodes allant jusqu'à 131 semaines. Un léger excès de fibrosarcomes a été noté dans la région des glandes salivaires et de la glande de Zymbal. Ces effets n'apparaissaient pas à dose plus faible (jusqu'à 35 g/m^3) et la dose la plus élevée n'a pas été utilisée lors de la seconde étude. Même si cela ne constitue pas une véritable démonstration de l'absence d'effets oncogènes, il n'y a pas eu d'augmentation de la fréquence des tumeurs après administration du composé par gavage à des rats. Ces animaux ont reçu le HCFC 22 à raison de 300 mg/kg par jour, 5 jours par semaine, pendant 52 semaines et l'étude s'est achevée au bout de 125 semaines.

7. Effets sur l'homme

On ne dispose que de données très limitées à propos des effets du HCFC 21 et du HCFC 22 sur l'homme.

Dans certains cas d'exposition accidentelle ou délibérée à de fortes concentrations de HCFC 22, les intéressés sont décédés. L'examen histopathologique d'échantillons de tissus prélevés sur quelques-unes de ces victimes a révélé la présence d'un oedème du poumon et d'inclusions lipidiques dans le cytoplasme, surtout dans les hépatocytes périphériques.

Un questionnaire que l'on a fait remplir par des personnes professionnellement exposées à du HCFC 22, a révélé une augmentation de l'incidence des palpitations, mais rien ne prouve avec certitude que l'exposition à l'un ou à l'autre des deux produits ait des effets nocifs. Une étude de mortalité a été effectuée sur des chlorofluorocarbures, dont le HCFC 22, mais elle portait sur un effectif trop faible pour qu'on puisse en tirer des conclusions.

8. Effets sur d'autres organismes au laboratoire et dans leur milieu naturel

On ne possède aucune donnée concernant les effets que le HCFC 21 ou le HCFC 22 pourraient exercer sur les êtres vivants dans leur milieu naturel.

RESUMEN

1. Identidad, propiedades físicas y químicas y métodos analíticos

Los dos clorofluorocarburos examinados en la presente monografía (diclorofluorometano, HCFC 21 y clorodifluorometano, HCFC 22) son hidroclorofluorocarburos (HCFC), es decir, compuestos obtenidos por la sustitución parcial de los átomos de hidrógeno del metano por átomos de flúor y cloro. Sólo el HCFC 22 tiene importancia comercial. El HCFC 21 y el HCFC 22 son gases no inflamables (a temperaturas y presiones normales), incoloros y prácticamente inodoros. El HCFC 21 es ligeramente soluble y el HCFC 22 moderadamente soluble en agua, y ambos son miscibles con disolventes orgánicos. El HCFC 22 puede encontrarse en forma de gas licuado.

Existen varios métodos analíticos para determinar esos dos compuestos. Entre ellos figuran la cromatografía de gases con captura electrónica y detección por ionización de llama, la cromatografía de gases/espectrometría de masas, y la espectrofotometría de deflección fototérmica.

2. Fuentes de exposición humana y ambiental

No se tiene noticia de que los dos HCFC estudiados en esta monografía existan como tales en la naturaleza. El HCFC 21 sólo se produce en pequeñas cantidades con fines no ocupacionales. Se calcula que la producción mundial total de HCFC 22 en 1987 ascendió a 246 000 toneladas.

La principal pérdida de HCFC 22 se produce durante la reparación, el uso y el desecho de frigoríficos y aparatos acondicionadores de aire. Actualmente se calcula que la pérdida mundial máxima está en torno a las 120 000 toneladas. Se han comunicado casos de liberación accidental de HCFC 22 en embarcaciones de pesca.

El HCFC 22 se utiliza como refrigerante como intermedio en la producción de tetrafluoroetileno, y como agente expansor del poliestireno. Una pequeña cantidad se utiliza como propulsor de aerosoles.

3. Transporte, distribución y transformación en el medio ambiente

El coeficiente de partición log octanol/agua del HCFC 22 es 1,08, lo que hace poco probable que se bioacumule. Se ha calculado que la persistencia del HCFC 21 en la troposfera es de unos 2 años y la del HCFC 22 de unos 17 años. Es probable que la reacción con radicales hidroxilo en la troposfera sea la principal vía de degradación. Sólo una pequeña fracción de los HCFC 21 y 22 alcanza la estratosfera donde, principalmente por reacción con radicales de oxígeno, liberan cloro que ataca al ozono. No obstante, se calcula que el HCFC 22 produce menos del 1% del cloro que ataca al ozono en la estratosfera. El potencial de destrucción de ozono (PDO) del HCFC 22 se ha calculado en 0,05 y se supone que el del HCFC 21 es aún menor.

Se estima que el potencial de calentamiento de la tierra (PCT), en relación con el del CFC 11, al que se asigna el valor 1,0, es inferior por un factor de 3 ó 4 el el caso del HCFC 22 y aún menor en el HCFC 21.

4. Niveles en el medio ambiente y exposición humana

No se dispone de datos sobre las concentraciones en el agua ni sobre la presencia de estos compuestos en los alimentos, aunque el HCFC 22 se utiliza en la fabricación de recipientes alimentarios de poliestireno expandido. No se tienen datos sobre la exposición humana al HCFC 21, pero en dos estudios sobre el uso de vaporizadores experimentales con un 17-65% de HCFC 22 se ha demostrado que las exposiciones breves (10-20 segundos) pueden originar concentraciones máximas que van desde 5000 hasta 8000 mg/m^3. Los trabajadores de peluquerías pueden estar expuestos a niveles medios ponderados en un tiempo de 8 horas de 90-125 mg/m^3, pero esos niveles se encuentran muy por debajo de los niveles reglamentarios MAK o LTV de 1800-3540 mg/m^3 en Alemania, EE.UU. y los Países Bajos.

El HCFC 22 se mezcla rápidamente en la atmósfera. En 1986 se comunicaron concentraciones de unos 326 mg/m^3 y se cree que el nivel aumenta en un 11% al año, aproximadamente.

5. Cinética y metabolismo en animales de laboratorio y en el ser humano

Se cuenta con datos limitados sobre la absorción, la distribución, el metabolismo y la excreción de HCFC 21. Puede inferirse que este compuesto se absorbe tras la inhalación a partir de los efectos sistémicos y de los elevados niveles de fluoruros en la orina observados en estudios de toxicidad en ratas. Las ratas exhalan HCFC 21 tras la inyección intraperitoneal y tanto los datos cinéticos como las pruebas de excreción de fluoruro sugieren que el HCFC 21 es metabolizado. No obstante, no se sabe hasta qué punto se metaboliza y, aparte del fluoruro, los productos no se han identificado.

En la rata, el conejo y el ser humano el HCFC 22 se absorbe rápidamente y sin dificultad tras la inhalación y se distribuye por todo el organismo. Se han encontrado niveles elevados de HCFC 22 en la sangre, el cerebro, el corazón, el pulmón, el hígado, el riñón y la grasa visceral de conejos moribundos durante la exposición y en muestras postmortem de cerebro, pulmón, hígado y riñón de víctimas accidentales de la exposición a HCFC 22. La eliminación es rápida; la mayor parte del HCFC se elimina con una semivida de 1 minuto en el conejo y 3 minutos en la rata. En el ser humano, una cantidad limitada de material se elimina en tres fases (semividas de 3 minutos, 12 minutos y 2,7 h).

El HCFC 22 inhalado o administrado por vía intraperitoneal se exhala casi por completo sin alteraciones tanto en la rata como en el hombre. Existen pruebas convincentes de que en la rata no se metaboliza en grado significativo ni *in vivo* ni en preparaciones de hígado.

6. Efectos en mamíferos de laboratorio y en sistemas de ensayo *in vitro*

No se dispone de datos satisfactorios sobre la toxicidad aguda por vía oral del HCFC 21 ni del HCFC 22.

Los efectos principales de una sola exposición por inhalación de HCFC 21 o HCFC 22 son esencialmente similares en diversas especies animales. Ambas sustancias tienen baja toxicidad por esta vía. Los efectos observados son característicos de los clorofluorocarburos, a saber, pérdida de coordinación y narcosis. Con elevadas concentraciones (106,7 g/m^3 o más) pueden producirse arritmias cardiacas y efectos pulmonares.

Aunque se ha indicado que tanto el HCFC 21 como el HCFC 22 causan irritación cutánea y ocular, esos efectos pueden haber guardado relación con las consecuencias de la pérdida de calor por evaporación antes que con las propiedades químicas de los HCFC. Ninguna de las dos sustancias provocó sensibilización cutánea.

Los únicos estudios realizados sobre la toxicidad a corto plazo del HCFC 21 han investigado la vía de inhalación. El principal efecto observado en la rata, el cobayo, el perro y el gato fueron las lesiones hepáticas; no se determinó el nivel sin efectos observados. Con niveles tan bajos como 0,213 g/m^3 administrados 6 h/día, 5 días por semana, durante 90 días, se observaron lesiones histopatológicas en el hígado. Con esa concentración también se observó edema pancreático intersticial y degeneración del epitelio de los túbulos seminíferos. En estudios realizados con el HCFC 22 a niveles de exposición entre 17,5 g por m^3 (durante 13 semanas) y 175 g/m^3 (durante 4 u 8 semanas) no se observaron lesiones.

No se han hecho estudios a largo plazo sobre el HCFC 21 en animales. La única conclusión coherente no tumorigénica observada en estudios a largo plazo con HCFC 22 fue la hiperactividad observada en ratones machos a los que se administraron 175 g/m^3, 5 h/día, 5 días por semana en un estudio de inhalación durante un lapso de vida entera.

No se han hecho estudios convencionales sobre los efectos del HCFC 21 en la fecundidad. En un estudio de la embriotoxicidad en ratas (42,7 g/m^3, 6 h/día los días 6-15 de la gestación) no se observaron efectos teratogénicos, pero se encontró una tasa elevada de pérdida de la implantación. El HCFC 22 (175 g/m^3 al día, 5 h/día, 5 días/semana durante 8 semanas) no tuvo efecto alguno en la capacidad de reproducción de las ratas macho. Al observarse un exceso pequeño, no significativo, de defectos oculares en tres estudios de teratología en ratas, se realizó un amplio estudio sobre la posible capacidad del

HCFC 22 de provocar defectos oculares. En ese estudio, se observó un aumento pequeño, aunque estadísticamente significativo, del número de camadas con fetos afectados de microftalmía o anoftalmía tras la exposición de la madre a 175 g/m^3, 6 h/día durante los días 6-15 de la gestación. Ese nivel de exposición produjo ligera toxicidad materna (peso corporal inferior en comparación con los testigos). No se observaron otros efectos; el nivel sin efectos observados en ese estudio fue de 3,5 g/m^3. El HCFC 22 no resultó teratogénico en un estudio convencional realizado en ratones con regímenes de exposición similares.

Se encontró que el HCFC 21 no era mutagénico en dos ensayos en bacterias y uno en levaduras (no se obtuvieron otros datos). El HCFC 22 fue mutagénico en ensayos bacterianos con *S. typhimurium*, pero no mostró actividad en ensayos con otros microorganismos ni en sistemas de mamíferos, ni *in vitro* ni *in vivo*. En esos ensayos se hicieron pruebas de mutación génica y de síntesis no programada del ADN *in vitro*, pruebas citogenéticas en médula ósea *in vivo*, y pruebas de letalidad dominante en la rata y en el ratón.

Sólo se han realizado ensayos de carcinogenicidad *in vivo* con el HCFC 22. Dos grupos de investigadores han realizado estudios de inhalación durante un lapso de vida entera tanto en ratas como en ratones. Tan sólo se observó excedente de tumores en el único estudio en el que se administraron a ratas macho 175 g/m^3, 5 días a la semana, durante hasta 131 semanas. Se observaron pequeños excedentes de fibrosarcomas en la región de las glándulas salivares y de la glándula de Zymbal. Esos efectos no se observaron con dosis inferiores (hasta 35 g/m^3), y esa dosis elevada no se utilizó en el segundo estudio. Aunque no sirva como demostración adecuada de la ausencia de efectos tumorigénicos, no se observó excedente de tumores en un estudio de alimentación forzada por vía oral en ratas. Se administró a esos animales HCFC 22 en dosis de 300 mg/kg al día, 5 días/semana, durante 52 semanas, y el estudio terminó a las 125 semanas.

7. Efectos en el ser humano

Se dispone de datos muy limitados sobre los efectos del HCFC 21 y el HCFC 22 en el ser humano.

Se han producido casos de muerte por exposición accidental o intencionada a niveles elevados de HCFC 22. El examen histopatológico de los tejidos de algunas de esas víctimas reveló la existencia de edema pulmonar y gotitas grasas citoplásmicas principalmente en los hepatocitos periféricos.

Aunque se ha indicado un aumento de la incidencia de palpitaciones en un estudio realizado con cuestionarios entre individuos expuestos al HCFC 22 en el trabajo, no existen pruebas sólidas de que la exposición de voluntarios o de trabajadores al HCFC 21 o al HCFC 22 produzca efectos nocivos en la salud. Nada puede concluirse a partir de un estudio de mortalidad sumamente reducido en personas expuestas por su profesión a varios clorofluorocarburos, entre ellos el HCFC 22.

8. Efectos en otros organismos en el laboratorio y sobre el terreno

No se dispone de datos sobre los efectos de los HCFC 21 y 22 en los organismos del medio ambiente.

www.ingramcontent.com/pod-product-compliance
Ingram Content Group UK Ltd.
Pitfield, Milton Keynes, MK11 3LW, UK
UKHW021308180426
11947UKWH00015B/1098